Impacts of Cross-Connections in North American Water Supplies

Impacts of Cross-Connections in North American Water Supplies

Prepared by:
J.J. Lee, **Paul Schwartz**, **Patrick Sylvester**, **Lawrence Crane**, **James Haw**, **Henry Chang**, and **H.J. Kwon**
Foundation for Cross-Connection Control and Hydraulic Research
University of Southern California
Kaprielian Hall 200
Los Angeles, CA 90089-2531

Sponsored by:
Awwa Research Foundation
6666 West Quincy Avenue
Denver, CO 80235-3098

Published by the

Awwa Research Foundation
Advancing the Science of Water®

IWA Publishing

DISCLAIMER

This study was funded by the Awwa Research Foundation (AwwaRF). AwwaRF assumes no responsibility for the content of the research study reported in this publication or for the opinions or statements of fact expressed in the report. The mention of trade names for commercial products does not represent or imply the approval or endorsement of AwwaRF. This report is presented solely for informational purposes.

Published by IWA Publishing, Alliance House, 12 Caxton Street, London SW1H 0QS, UK

Telephone: +44 (0) 20 7654 5500; Fax: +44 (0) 20 7654 5555; Email: publications@iwap.co.uk;
Web: **www.iwapublishing.com**

AwwaRF report number 90928F. Originally published by AwwaRF for its subscribers in 2003.
IWA Publishing version published 2004.

ISBN 1 84339 843 5

CONTENTS

LIST OF TABLES

LIST OF FIGURES

FOREWORD

The Awwa Research Foundation is a nonprofit corporation that is dedicated to the implementation of a research effort to help utilities respond to regulatory requirements and traditional high-priority concerns of the industry. The research agenda is developed through a process of consultation with subscribers and drinking water professionals. Under the umbrella of a Strategic Research Plan, the Research Advisory Council prioritizes the suggested projects based upon current and future needs, applicability, and past work; the recommendations are forwarded to the Board of Trustees for final selection. The foundation also sponsors research projects through the unsolicited proposal process; the Collaborative Research, Research Applications, and Tailored Collaboration programs; and various joint research efforts with organizations such as the U.S. Environmental Protection Agency, the U.S. Bureau of Reclamation, and the Association of California Water Agencies.

This publication is a result of one of these sponsored studies, and it is hoped that its findings will be applied in communities throughout the world. The following report serves not only as a means of communicating the results of the water industry's centralized research program but also as a tool to enlist the further support of the nonmember utilities and individuals.

Projects are managed closely from their inception to the final, report by the foundation's staff and large cadre of volunteers who willingly contribute their time and expertise. The foundation serves a planning and management function and awards contracts to other institutions such as water utilities, universities, and engineering firms. The funding for this research effort comes primarily from the Subscription Program, through which water utilities subscribe to the research program and make an annual payment proportionate to the volume of water they deliver and consultants and manufacturers subscribe based on their annual billings. The program offers a cost-effective and fair method for funding research in the public interest.

A broad spectrum of water supply issues is addressed by the foundation's research agenda: resources, treatment and operations, distribution and storage, water quality and analysis, toxicology, economics, and management. The ultimate purpose of the coordinated effort is to assist water suppliers to provide the highest possible quality of water economically and reliably. The true benefits are realized when the results are implemented at the utility level. The foundation's trustees are pleased to offer this publication as a contribution toward that end.

Edmund G. Archuleta, P.E.
Chair, Board of Trustees
Awwa Research Foundation

James F. Manwaring, P.E.
Executive Director
Awwa Research Foundation

ACKNOWLEDGMENTS

The Foundation for Cross-Connection Control and Hydraulic Research at the University of Southern California wishes to acknowledge the Project Advisory Committee, Craig H. Adams from Tucson Water, Jung J. Choi from the Philadelphia Water Department, and Randy Pankiewicz from the Pennsylvania-American Water Company for their guidance and comments throughout the project duration. The Foundation would also like to acknowledge Stephanie Morales for her leadership and advice in every aspect of this project. The Foundation also acknowledges Scott Foster for his help in the modeling. A special thanks is also given to the participating water utilities, which contributed manpower to this project.

EXECUTIVE SUMMARY

A comprehensive assessment of cross-connections was conducted in this study. Over 700 water utilities participated in this study. The study consisted of two major components. A comprehensive survey with an extensive relational database system was created to ascertain various issues related to cross-connection. Computer simulations were conducted for three different sized water distribution systems, ranging from large to small. Several scenarios were simulated to study the transient effects of rapid changes in demand.

SUMMARY AND CONCLUSIONS

No geographic trends were found in North America in relation to the type of cross-connections being experienced. Indirect cross-connections account for 65% of the cross-connections while 35% are direct cross-connections.

The North American survey data indicates that more and more agencies developed programs over the last three decades. As more agencies developed programs, more incidents were reported. Therefore, the reporting of incidents has continually increased since 1975. It is important, however, to note that the average number of incidents being reported per agency does not significantly change. This suggests that the number of incidents per utility isn't necessarily changing over time, but that reporting is becoming more prevalent as agencies develop cross-connection control programs.

The overall average cost for all sizes of water utilities is $44,835 spent annually on administering the cross-connection control program. On the average 3.65% of the agencies' operations and maintenance budget was directed towards cross-connection control.

The data analysis demonstrated that those utilities with higher maximum elevation differences experienced a greater number of backflow incidents. Analyzing a combination of nominal system pressure and maximum elevation difference indicated that systems with lower nominal pressure and higher elevation differences experienced more incidents than systems with higher nominal pressure and lower elevation differences. This conclusion is also demonstrated in the computer modeling.

Ninety-one percent of the utilities responding to the North American survey indicated that they did have a cross-connection control program. The survey data shows that 43% of the utilities surveyed administered system protection programs (their goal is to protect the water system at the users water connection).

Establishing a legal authority was ranked as the most effective element of a cross-connection control program. Having a list of approved backflow preventers and then a requirement to test the backflow preventers followed this.

Results from the computer modeling clearly indicate that sudden changes in the water demand, such as a hydrant opening, caused drastic transient flow conditions resulting in significant pressure loss and flow reversal in a network system of any size. Many examples demonstrate that the loss in pressure could reach below atmospheric pressure.

Distribution systems that have greater variance in ground elevations will have greater possibility of creating sub-atmospheric pressure when drastic pressure reduction occurs.

Slower hydrant opening rates significantly reduce the impact of hydraulic transients created.

The extent of the affected area cannot be generalized. The affected area depends on the specific condition creating sudden changes and the particular system characteristics of the network.

RECOMMENDATIONS TO WATER UTILITIES

Water utilities should initiate and/or maintain cross-connection control programs, which include:

1. A legal document governing the program
2. The use of listed or approved backflow prevention assemblies
3. Field-testing the backflow prevention assemblies
4. Trained, licensed, or certified personnel to field-test the backflow preventers
5. Site surveys to determine the extent of backflow prevention required
6. A requirement to keep records and document backflow incidents
7. A public education effort to both the general public and water utility personnel

The computer modeling provided clear evidence that sudden changes in demand can clearly cause undesirable conditions within the distribution system. Appropriate plans to eliminate the reverse flow of water and thereby backflow should be made prior to potential public health and financial disasters. These sudden changes may be due to main breaks, fire hydrant use, or other changes in water demand. The impact of rapid pressure drop and flow reversal could be lessened if the rate of change is decreased. Although this may not be possible in the case of a main break, it is possible in cases of flushing hydrants and other controlled changes in the system. Operations personnel need to be trained to make these changes gradually in order to lessen the effect of rapid system changes. The computer modeling, along with the survey results, showed that systems with larger elevation differences are more susceptible to the negative impact of sudden changes of water demand in the system.

Since the evidence shows that sudden changes in demand can clearly create undesirable conditions within the distribution system, appropriate plans should be made prior to potential disasters. These plans can only be made after a careful study of the response of the system to various assumed scenarios.

CHAPTER 1
INTRODUCTION

BACKGROUND

One of the most significant engineering feats in the history of civilization has been the advent of potable water distribution networks. However, with this comes the solemn responsibility of safeguarding the integrity of the potable distribution system.

Technical literature and the public press have recorded incidents of backflow of contaminated or polluted substances into the distribution systems of water utilities for well over 100 years. A few water utilities instituted modest cross-connection control programs in the 1930's. However, only recently has there been a steady increase in the number of cross-connection control programs across North America.

Although many cross-connection control programs have been initiated in recent decades, there are many utilities, which either do not have a program or do not have a sufficient program to provide for reasonable protection from cross-connections. Part of this is due to a lack of uniform regulations. Another reason many utilities have not initiated or maintained strong cross-connection control programs is that vast quantities of documented data regarding cross-connections are not readily available. This is mostly due to the nature of backflow.

Hydraulic conditions within a piping system change continuously. As a brief example, it is possible for a water main break to cause a drop in system pressure within a region of a water distribution system. Because of cross-connections common to any distribution system, backflow could occur contaminating a portion of the distribution system. As a result of this backflow situation, individuals could ingest the contaminant, or the appearance of water could cause alarm to the water user. Before the source of the contaminant can be discovered, the water main is repaired, water pressure is normalized and the system flushed clean through normal use. This makes it very difficult to document such a backflow incident.

Since there is no Federal reporting requirement for cross-connections, national statistics are not available. There are, however, a number of backflow incidents, which have been compiled by the Foundation for Cross-Connection Control and Hydraulic Research at the University of Southern California. Selected portions of this information is published in the Manual of Cross-Connection Control, Ninth Edition. This is by no means all-inclusive, but does give an indication that cross-connections are widespread and nationwide.

Any substance that can come into contact with the potable water distribution system may cause pollution or contamination. The substance, or substances, may be introduced into the potable water system by a condition called backflow, which by definition is the undesirable reversal of flow. Should a substance backflow into the potable water system, it may present an acute threat (short term exposure) or a chronic threat (long term intermittent exposure) to the water quality. Through the Safe Drinking Water Act (SDWA), Congress requires the EPA[1] to regulate contaminants which may be health risks and which may be present in public drinking water supplies. Passed in 1974, and revised in 1986 and 1996, the SDWA is intended to provide health protection to the consumers of drinking water. The SDWA and regulations developed under it are extensive. Monitoring and reporting requirements are specified in the regulations.

---

[1] USEPA. 1994. *Drinking Water Regulations and Health Advisories.* Office of Water. Washington D.C.

Monitoring of drinking water ranges from naturally occurring chemicals to man made contaminants.

Examples of substances that have backflowed into the potable water distribution system include:

- Pathogenic microorganisms – fecal coliforms, giardia
- Inorganic compounds – nitrates (fertilizers)
- Pesticides – chloradane, diazinon
- Metals – lead, chromium, copper
- Radionuclides – uranium

Once any of these contaminants enter the distribution system through a cross-connection, there is no way to predict how it will react. A number of parameters have to be considered:

- Contaminant/Pollutant
- Chemical/physical properties of the material
- Distribution system
    - Point of Entry
        - Consumer's plumbing fixture(s)
        - Connection to public water system
    - Line pressure
    - Rate of flow

Due to the many variables in the distribution system, the contaminant may be diluted, thus reducing the concentration of the pollution or contamination, or the material may maintain a high concentration in a slug form.

Water distribution systems have always been susceptible to cross-connections and backflow, but the recognition of a backflow incident was not widely understood. Backflow incidents have been suspected, and some recorded, for as long as there have been formal water distribution systems. In earlier days, contamination of the drinking water system has mainly been attributed to problems with the source water (groundwater or surface water supplies), and not necessarily with the distribution system itself. However, in the book *Waterborne Diseases in the United States*, by Gunther F. Craun, the importance of cross-connections is recognized. Craun compiled and examined all of the available data on waterborne outbreaks from 1920-1981 to provide a historical perspective of the epidemiology of waterborne disease. Of the waterborne outbreaks attributed to distribution systems from 1920 - 1980, it states; "Cross-connections and backsiphonage were responsible for most (78%) outbreaks and disease (95%) caused by distribution system contamination."[2]

---

[2] Craun, Gunther F. 1986. *Waterborne Diseases in the United States*. CRC Press, Inc. pp. 131, 134.

Lack of Documentation

The documentation of backflow incidents has been limited at best. The USEPA Office of Inspector General stated in a 1995 report [3] "...contamination of the potable water supply by cross-connections is largely undetected, not investigated, not properly documented, or not reported." This conclusion was reached from interviews with various industry experts.

The underreporting of backflow incidents may be attributed to a number of factors. These factors will be addressed as follows:

To trace the origin of the pollution or contamination of the drinking water system may be extremely difficult. Due to the transient nature of backflow, a cross-connection may exist for many years and nothing occurs. However, with the correct set of conditions occurring (i.e., loss of supply pressure, process or equipment in operation, etc.) the pollution or contamination may be forced back into the drinking water. Once the proper or normal operating conditions return, the backflow of the substance stops.

Routine bacteriological monitoring may not detect the contamination, especially in small water systems with a limited number of routine samples.[4] Once again, due to the short-lived transient nature of backflow, a backflow occurrence that takes place between the routine sampling would go undetected (sampling only provides a snapshot in time of the water quality at a specific location). Should a routine bacteriological sample indicate a contamination, a repeat sample may not detect the contamination as still being present.

Chemical contamination may go undetected by routine bacteriological monitoring.

Waterborne chemical contaminants are probably underreported to CDC. Several reasons could explain the low reporting rate, including the following[5]:

Most incidents of this nature probably occur in private residences, affect relatively few persons, and might not come to the attention of public health officials.

Exposure to chemicals via drinking water can a) cause illness that is difficult to attribute to chemical intoxication or b) cause nonspecific symptoms that are difficult to link to a specific chemical.

The mechanisms for detecting waterborne chemical contaminants and reporting them to the WBDO (waterborne-disease outbreak) surveillance system are not as well established as those for WBDO's caused by infectious agents.

Physicians might have difficulties recognizing and diagnosing chemically caused symptoms.

The reporting of WBDO's to CDC is voluntary. State, territorial, and local public health agencies have the primary responsibility for detecting and investigating WBDO's.[6]

Current Documentation

For many of the reasons identified above, proper recognition and documentation of backflow incidents has been limited. Documentation of backflow incidents has been compiled

---

[3] USEPA. 1995. *Survey Report on the Cross-Connections Control Program.* Office of Inspector General. p. 10.
[4] ABPA. 1999. *Survey of State and Public Water System Cross-Connection Control Programs.* p. 4.
[5] CDC. 2000. "Surveillance for Waterborne-Disease Outbreaks – United States, 1997-1998." *Morbidity and Mortality Weekly Report 49.* p. 12.
[6] CDC. 1998. "Surveillance for Waterborne-Disease Outbreaks – United States, 1995-1996." *Morbidity and Mortality Weekly Report 47.* p. 3.

by several organizations, but there has not been one central repository. This documentation has typically been supplied by the water or health agency in the affected area, or interested parties learning about the incident in the media (i.e., newspapers, periodicals, radio, television). Some of the organizations involved with compiling this data are:

- American Backflow Prevention Association (ABPA)
- Pacific Northwest Section of the American Water Works Association (PNWS-AWWA)
- University of Southern California - Foundation for Cross-Connection Control and Hydraulic Research (USC-FCCCHR)
- Centers for Disease Control (CDC)
- United States Environmental Protection Agency (USEPA)

Starting in 1971, CDC–in collaboration with the USEPA–started compiling data on waterborne disease outbreaks in annual surveillance reports. Some of the goals of this reporting included[7]:

- to determine trends in the incidence of water related diseases in the United States
- to disseminate information on prevention and control of water related diseases to public health personnel

In 1999, the EPA began to compile data in this area as part of the work done for the Stage 2 Microbial and Disinfection Byproducts Federal Advisory Committee (FACA). The purpose of the Committee was to evaluate whether and to what degree the USEPA should establish revised or additional disinfection byproducts and microbial standards to protect public health. The Agreement in Principle[8] was signed in September 2000 by the FACA members, with the following statements in Part B (all but one member of the FACA signed Part B):

7.3 The FACA recognizes that cross connections and backflow in distribution systems represent a significant public health risk

7.5 Therefore, the FACA recommends that beginning in January 2001, as part of the 6-year review of the Total Coliform Rule, EPA should evaluate available data and research on aspects of distribution systems that may create risks to public health and, working with stakeholders, initiate a process for addressing cross connection control and backflow prevention requirements and consider additional distribution system requirements related to significant health risks.

MEANS TO PREVENT BACKFLOW

Where cross-connections can not be eliminated, then the installation of a backflow preventer may be necessary. A backflow preventer is a mechanical assembly, device or method that prevents the unwanted reversal of flow (i.e., backflow). The type of backflow preventer utilized will depend upon the type of cross-connection (i.e., direct or indirect); the degree of hazard (i.e., non-health, health, or sewage); and if the application is under continuous pressure or

---

[7] Waterborne-Disease Outbreaks. 1990. *Morbidity and Mortality Weekly Report*, United States. pp. 39, 1.
[8] US Federal Register. 2000. *Volume 65*. pp. 83015-83024.

usage. Some of the more common backflow preventers are listed below followed by a brief description of each:

1. Air Gap
2. Vacuum Breakers
3. Reduced Pressure Principle Backflow Prevention Assemblies
4. Double Check Valve Backflow Prevention Assemblies

Air Gap

The term "air gap" shall mean a physical separation between the free flowing discharge end of a potable water supply pipeline and an open or non-pressure receiving vessel. An "approved air gap" shall be at least double the diameter of the supply pipe measured vertically above the overflow rim of the vessel — in no case less than 1 inch (2.54 cm). ASME A112.1.2 - 1991 *Air Gaps in Plumbing Systems*[9].

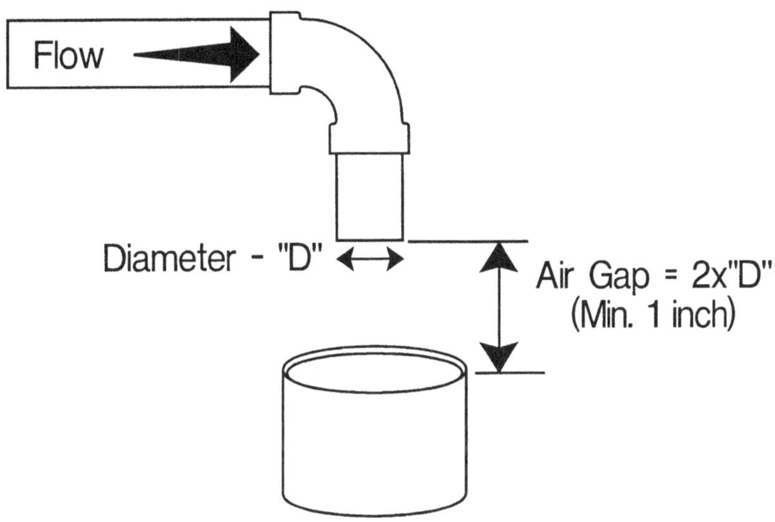

Figure 1.1 Air Gap

Vacuum Breakers

Atmospheric Vacuum Breaker Backsiphonage Prevention Assembly (AVB)

The term "atmospheric vacuum breaker backsiphonage prevention assembly" or AVB (also known as the non-pressure type vacuum breaker) shall mean an assembly containing an air inlet valve, a check seat and an air inlet port(s). The flow of water into the body causes the air inlet valve to close the air inlet port(s). When the flow of water stops the air inlet valve falls and forms a check valve against backsiphonage. At the same time it opens the air inlet port(s) allowing air to enter and satisfy the vacuum. A shutoff valve immediately upstream may be an integral part of the assembly, but the assembly shall not be subjected to operating pressure for

[9] ASME A112.1.2 - 1991 *Air Gaps in Plumbing Systems*

more than twelve (12) hours in any twenty-four (24) hour period. An atmospheric vacuum breaker is designed to protect against a non-health hazard (i.e., pollutant) or a health hazard (i.e., contaminant) under a *backsiphonage* condition only. [10]

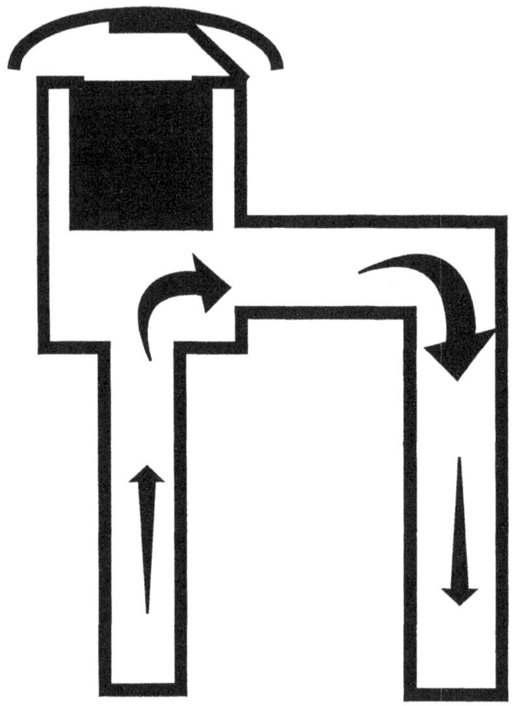

Figure 1.2 Atmospheric Vacuum Breaker Backsiphonage Prevention Assembly

Hose Connection Vacuum Breakers

Hose Connection Vacuum Breakers (ASSE Standard #1011)[11] are designed to be installed on the discharge side of a hose bib. When not under pressure, the checking member is loaded to the closed position, while the atmospheric port(s) are loaded to the open position. This device is intended for non-continuous use (i.e., 12 hours use in 24 hour period).

Pressure Vacuum Breakers

The term "pressure vacuum breaker backsiphonage prevention assembly" shall mean an assembly containing an independently operating internally loaded check valve and an independently operating loaded air inlet valve located on the discharge side of the check valve. The assembly is to be equipped with properly located resilient seated test cocks and tightly

---

[10] USC. 1993. *Manual of Cross-Connection Control 9th Edition*. University of Southern California.

[11] ASSE Standard No. 1011. 1993. *Performance Requirements for Hose Connection Vacuum Breakers*. American Society of Sanitary Engineering.

closing resilient seated fully ported shutoff valves attached at each end of the assembly. This assembly is designed to protect against a non-health hazard (i.e., pollutant) or a health hazard (i.e., contaminant) under a backsiphonage condition only.

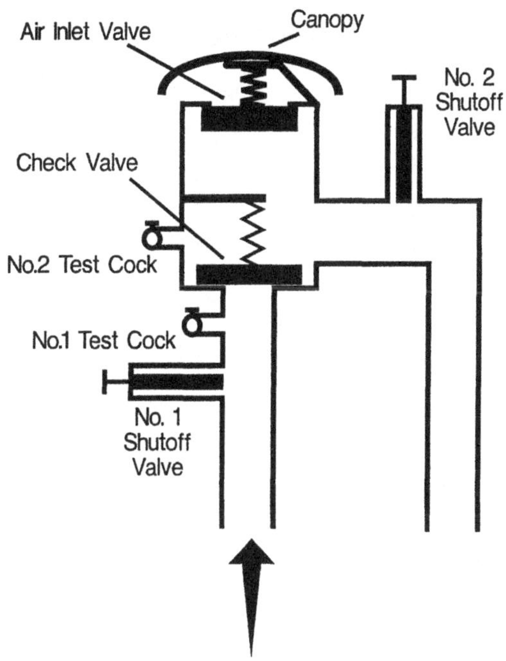

Figure 1.3 Pressure Vacuum Breaker Backsiphonage Prevention Assembly

Spill-resistant Pressure Vacuum Breakers

The term "spill-resistant pressure vacuum breaker backsiphonage prevention assembly" shall mean an assembly containing an independently operating internally loaded check valve and independently operating loaded air inlet valve located on the discharge side of the check valve. The assembly is to be equipped with a properly located resilient seated test cock, a properly located bleed/vent port, and tightly closing resilient seated fully ported shutoff valves attached at each end of the assembly. This assembly is designed to protect against a non-health hazard (i.e., pollutant) or a health hazard (i.e., contaminant) under a backsiphonage condition only.

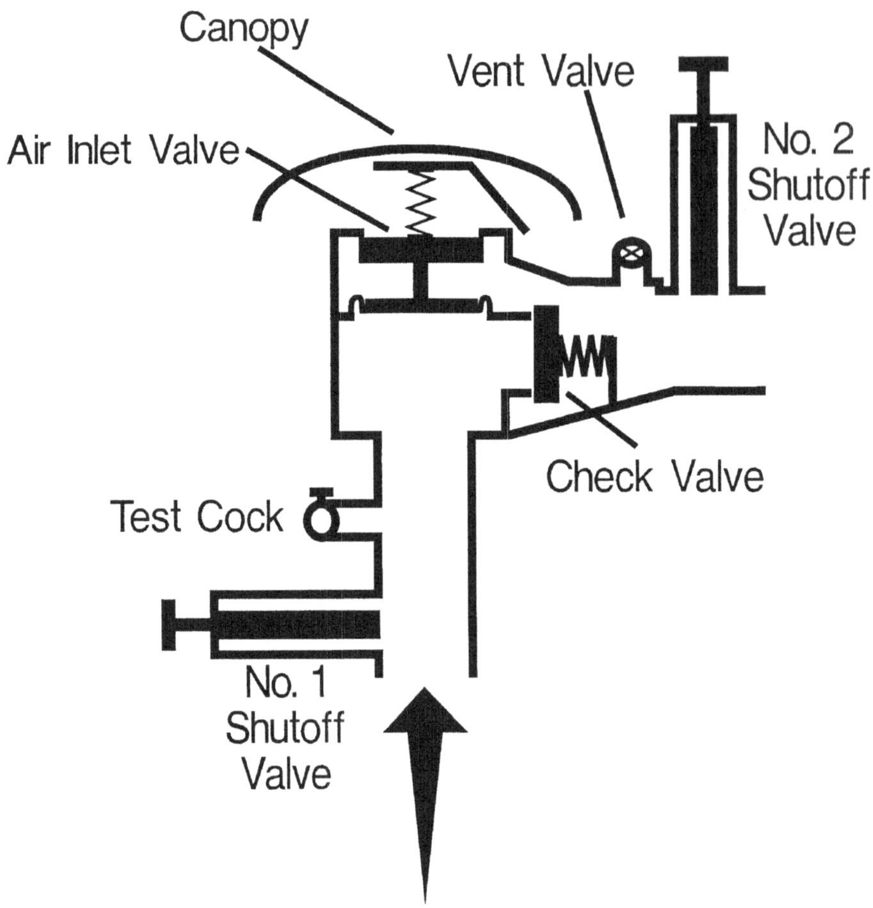

Figure 1.4 Spill-Resistant Pressure Vacuum Breaker Backsiphonage Prevention Assembly

Reduced Pressure Principle Backflow Prevention Assemblies

The term "reduced pressure principle backflow prevention assembly" shall mean an assembly containing two independently acting approved check valves together with a hydraulically operating, mechanically independent pressure differential relief valve located between the check valves and at the same time below the first check valve. The unit shall include properly located resilient seated test cocks and tightly closing resilient seated fully ported shutoff valves at each end of the assembly. This assembly is designed to protect against a non-health (i.e., pollutant) or a health hazard (i.e., contaminant). This assembly shall not be used for backflow protection of sewage or reclaimed water. (Note: Check with local administrative authority for acceptable uses.)

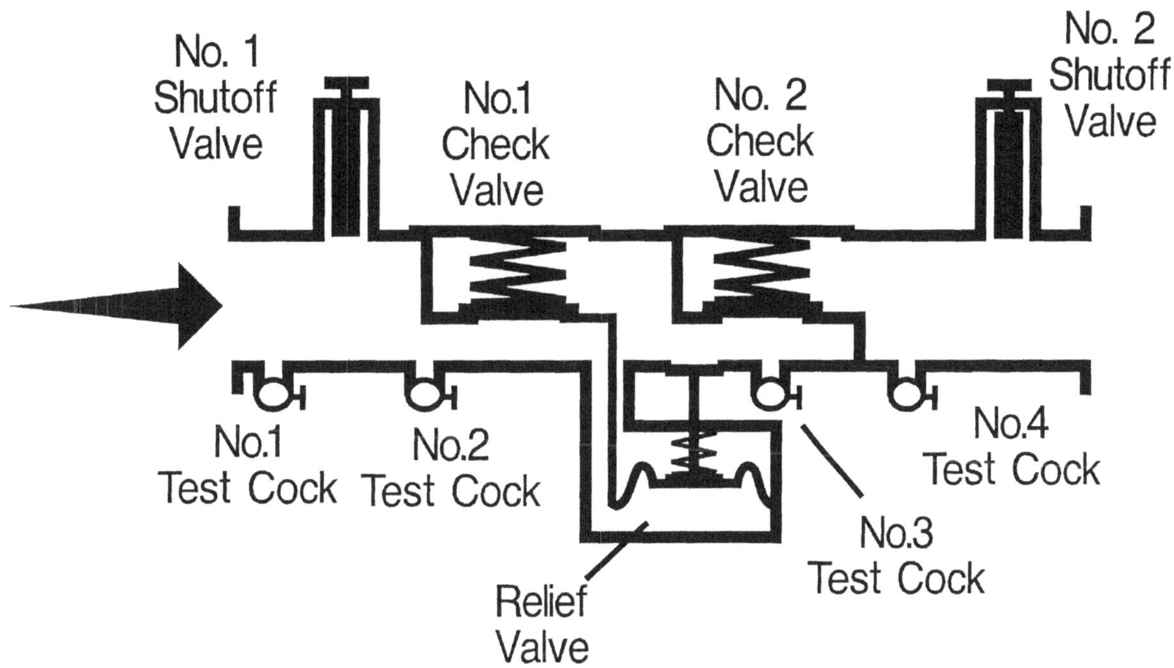

Figure 1.5 Reduced Pressure Principle Backflow Prevention Assembly

Double Check Valve Backflow Prevention Assemblies

The term "double check valve backflow prevention assembly" shall mean an assembly composed of two independently acting, approved check valves, including tightly closing resilient seated fully ported shutoff valves attached at each end of the assembly and fitted with properly located resilient seated test cocks. This assembly shall only be used to protect against a non-health hazard (i.e., pollutant).

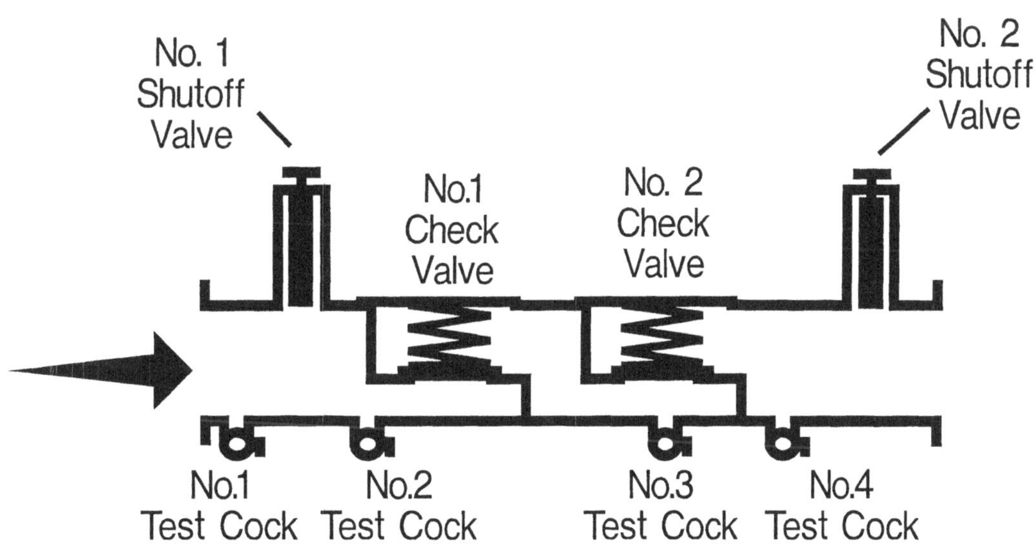

Figure 1.6 Double Check Valve Backflow Prevention Assembly

Other Devices

Dual Check (ASSE Standard #1024)[12] – These devices contain two internally loaded check valves. These devices are intended to protect the potable water supply from non-health hazards at residential service lines and individual outlets.

Dual Check with Atmospheric Port (ASSE Standard #1012)[13] – These devices contain two independently operating loaded check valves separated by an intermediate chamber with a means for venting to the atmosphere. These devices are designed to give protection against low hazard backpressure and/or backsiphonage conditions on individual outlets. They are not designed nor intended for building isolation or high hazard conditions.

Single Check Valves –A single check valve may be biased or loaded to the closed position, but common single check valves used in plumbing systems do not contain a loading member (i.e., spring or weight loaded). In addition, these check valves typically contain metal-to-metal sealing surfaces, which are not permitted in backflow prevention assemblies. Single check valves of any type have not been and should not be considered backflow preventers, as described above, since they do not contain a redundant protection member. If the single check valve is fouled or leaks, there is no secondary or backup protection.

---

[12] ASSE Standard No. 1024. 1994. *Performance Requirements for Dual Check Valve Type Backflow Preventers.* American Society of Sanitary Engineering.

[13] ASSE Standard No. 1012. 1993. *Performance Requirements for Backflow Preventer with Intermediate Atmospheric Vent.* American Society of Sanitary Engineering.

[13] ASSE Standard No. 1012. 1993. *Performance Requirements for Backflow Preventer with Intermediate Atmospheric Vent.* American Society of Sanitary Engineering.

CHAPTER 2
METHODS AND MATERIALS

THE NATIONAL SURVEY AND WORKSHOPS

In order to obtain a valid assessment of cross-connections in North American water supplies it was necessary to obtain primary data. Secondary data was available for little of the information needed. It was, therefore, necessary to conduct a survey of North American water supplies to obtain primary data regarding cross-connections. The survey was conducted in the form of a questionnaire. This questionnaire was designed to obtain a great deal of information, and yet not be such a burden to the respondent as to discourage completion. To encourage completion of the questionnaire, the recipients were provided a business reply envelope and were told they would be invited to a workshop discussing the findings of the project at a later date. Wherever possible the survey questions included multiple-choice answers, in order to simplify the completion process. There were, however, some questions that required information to be filled in. Additionally, some of the questions required the responder to conduct a fair amount of research before being able to complete the questions accurately. The intent of the survey was to gather information on the eleven points listed below. Some of the points listed below were combined together in the *Results and Discussion* portion of the report. This is due to the fact that much of the information overlapped into adjacent areas. In addition to the survey and the follow up survey, two workshops were conducted to gain more information. Both workshops were well attended and offered the research team a large amount of general information through the commentary expressed in response to the information presented at each workshop.

Most Common Types of Cross-Connections

A cross-connection is defined as follows in the *Manual of Cross-Connection Control*, Ninth Edition:

The term "cross-connection" shall mean any unprotected actual or potential connection or structural arrangement between a public or a consumer's potable water system and any other source or system through which it is possible to introduce into any part of the potable system any used water, industrial fluid, gas, or substance other than the intended potable water with which the system is supplied. Bypass arrangements, jumper connections, removable sections, swivel or change-over devices and other temporary or permanent devices through which or because of which backflow can occur are considered to be cross-connections.

a. The term "direct cross-connection" shall mean a cross-connection which is subject to both backsiphonage and backpressure.

b. The term "indirect cross-connection" shall mean a cross-connection which is subject to backsiphonage only.

The survey was designed to determine which types of cross-connections (i.e., direct, or indirect) are most prevalent in distribution systems as well as within user premises. The analysis of this data will determine if specific types of cross-connections are more prevalent in specific geographic regions, or types of systems. Additionally, the types of cross-connections most commonly found in distribution systems will be compared and contrasted with types of cross-connections most commonly found in internal on-premise water systems.

History of Backflow Incidents

A comprehensive analysis of data provided from the survey was designed to determine the history of backflow incidents, which have occurred in water distribution systems and on-premise systems. Determining if there is a correlation between backflow incidents and efforts on behalf of the administrative authority to abate cross-connections is part of the analysis.

Financial Impacts

The financial impacts of backflow incidents are analyzed including incident response, investigation, clean up and repair.

Hydraulic Changes within a Distribution System

Determination of the hydraulic changes within a water distribution system and the changes' potential to create backflow conditions is made. Changes in system pressures change the direction of the flow of water, by varying high and low pressure regions within a system. A comprehensive analysis of how this affects the potential for backflow is conducted. This information is compared with the historical data on backflow incidents to identify patterns of backflow occurrence under certain hydraulic conditions within the system. This information will help administrative authorities prioritize their backflow prevention program based on hydraulic conditions in the system.

Assessment of the Occurrence of Backpressure and Backsiphonage

An assessment of the data obtained determines if there is any correlation between the type of backflow, which occurs (i.e., backpressure or backsiphonage) and various water systems and communities.

Types of Cross-Connection Control Programs

A detailed analysis of cross-connection control programs determines the major elements of an effective cross-connection control program. The types of programs in effect, combined with other elements of the survey show which specific aspects of a program provide the greatest public health benefit.. This information also determines if specific aspects, such as public education are helpful in bringing about compliance.

An analysis of the effectiveness of internal protection verses system protection is made.

Regulations and Enforcement Issues

The common aspects of the most effective regulations are isolated from the survey data. The most effective means of enforcement is determined. Whether or not a public relations program assists in the overall enforcement of a program is assessed as well as what effect public education has on the enforcement capability of a program. The effectiveness of guidance manuals, specific policies and procedures are also assessed.

Elements of a Cross-Connection Control Program

The most effective elements of a cross-connection control program are identified. Included are:

- What type of legal authority is effective in a cross-connection control program?
- What elements of the regulation, or ordinance are necessary, most effective and enforceable?
- What types backflow preventers are most effective?
- Are certain types of backflow preventers most effective under specific conditions?
- Is there a regular field-testing requirement for backflow preventers?
- What methods are effective for testing the backflow preventers?
- What requirements are needed for those field-testing the backflow preventers?
- What requirements are needed for those surveying sites to determine what, if any, backflow prevention is needed on the premises?
- What types of cross-connection records are maintained and how detailed are they?
- How important/effective is education in the program?

Additionally, other elements of an effective program are identified via the survey.

Relationship of Public Education to Enforcement

The survey assesses the effectiveness of public education as it relates to enforcement and administration of the cross-connection control program.

Evaluation of Customer Complaints

The survey determines the probability of specific customer complaints being related to cross-connection incidents. An analysis of the number and type of complaints are associated with common cross-connections to see if there is a correlation.

Administration of Cross-Connection Control Programs

Common problems encountered in administering a cross-connection control program are analyzed to determine if there are specific recurring problems in different regions of North America. An analysis of how administrative authorities work together to abate cross-connections is conducted.

METHODOLOGY OF DATA CORRELATION

The initial Survey shown in Appendix A was sent to 4449 Water Utilities. Of these, 719 (or 16%) responded. As this data was being compiled, it became evident that additional information would be required. Therefore, with the input of the Project Advisory Committee

(PAC), and after much deliberation, the follow-up survey, as shown in Appendix B was sent to the 719 respondents to the initial survey. Of original respondents, 352 (or 49%) responded to the follow up survey.

In order to ensure consistency, one individual handled the data entry. Random checks by other members of the research team were conducted in order to verify accuracy of the entries. Once all of the data was entered, summaries of the results were printed to allow the research team an overview of the results. This was used in determining which data correlations would be run. As the summary of the initial survey was reviewed, it became evident that the follow-up survey would be necessary in order to clarify some of the responses and gather more detailed information. The data from both of the surveys was input into a Microsoft ACCESS database. This gave the researchers the flexibility to create a variety of queries to determine correlations between different answers on both surveys. Much of the needed information came directly from responses to questions on the surveys. Some of the information, however, could only be determined by running correlations on the survey results. The research team met to discuss which correlations would provide the best information relating to the points listed above. The correlations were run in order to determine the specific information generated in the body of this report.

CHAPTER 3
RESULTS AND DISCUSSION OF THE NORTH AMERICAN SURVEY

OVERVIEW OF THE RESPONDING UTILITIES

There were two surveys sent to water utilities for the project. Seven hundred nineteen respondents completed the initial survey, at least in part. All 719 respondents did not answer every question. To clarify some of the responses received from the initial 719 respondents, a second follow-up survey was mailed. Almost one-half of the initial respondents responded to the follow-up survey (352). Responses were received from forty-six States and the District of Columbia of the United States; and, five Canadian Provinces.

Of those responding to the survey, 99% identified their utility as a community water system. Ninety percent were publicly owned and 10% privately owned. Wholesalers accounted for 29.8%, 62.2% provided water as a retailer, 4.82% to agricultural systems, and 3.2% provided water in another way. Eight percent of the respondents indicated having less than 500 service connections; 11%, 500 to 1000; 36%, 1001 to 5000; 30%, 5001 to 25,000; 11%, 25,001 to 100,000 and 4% indicated having more than 100,000 service connections.

Recycled water systems were operated by 10.7% of the respondents. Backflow protection was required by the 52.3% of the utilities in these recycled water systems. Seventy-two percent of these stated the type of protection required was a reduced pressure principle assembly; 9.1% indicated a double check valve assembly was used for protection on these systems; 12.1% indicated "other," most of whom wrote in "air gap."

The respondents were asked to provide the number of service connections in each of four categories. On the average, the responses indicated that 86% of the service connections were residential domestic; 9%, commercial/industrial domestic; 3%, fire suppression and 2%, irrigation systems. The respondents were requested to determine how many connections in each of these categories had backflow protection. The results showed that 30.5% of commercial/industrial domestic systems, 7.9% of residential connections, 46% of irrigation systems, and 53% of fire suppression systems were provided with backflow protection.

Additional information obtained from the survey informed the researchers that 94% of the respondents' service connections were metered. Of those service connections, which were not metered, 64% were fire protection systems; 13%, residential; 11%, construction, and 12% indicated "other." Those who marked "other" wrote in types of services such as town or city facilities, hydrants, public, irrigation, parks and a few miscellaneous connections. Those surveyed were also asked to provide the sizes of their service connections. The results are shown in Table 3.1.

Table 3.1
Distribution of size of connection in the system

Size of Connection	% of total system
½"	2.6
¾"	14.8
1"	15.3
1 ½"	12.1
2"	14.3
3"	8.5
4"	10.4
6"	9.6
8"	5.8
10"	3.4
12"	1.6
Other*	1.5

*The majority of those who marked "other" wrote in 5/8-inch.

MOST COMMON TYPES OF CROSS-CONNECTIONS

One of the goals of the research was to determine the extent of direct versus indirect cross-connections. Direct cross-connections are those subject to backpressure or backsiphonage. Indirect cross-connections are those subject to backsiphonage only. It was important to determine the extent of indirect vs. direct cross-connections since there are different means of protecting against each type.

According to the tabulated survey results the national average was found to be 65% indirect to 35% direct cross-connections. An analysis was made to determine if there are certain trends across North America where geographic areas might be experiencing significantly higher or lower averages when compared to the national average. In analyzing the survey data, states that differed by more than 15% from the national average were highlighted. No geographic trends were found in North America in relation to the type of cross-connections being experienced. For example, while Washington, DC experienced more than 15% lower rates of indirect cross-connections, Pennsylvania, in the same geographic region, experienced more than 15% higher rates of indirect cross-connections when compared to the national average. The same situation occurred in the South with Alabama and Louisiana and the same in the Midwest with Kansas and Illinois. The types of cross-connections by state are shown in Table 3.2.

Table 3.2.
Type of Cross-Connections by state

State/Province	% Indirect Cross-Connections	% Direct Cross-Connections	State/Province	% Indirect Cross-Connections	% Direct Cross-Connections
Alberta	20	80	Mississippi	60	40
Alaska	95	5	Montana	53	47
Alabama	50	50	New Brunswick	53	47
Arkansas	70	30	North Carolina	72	28
Arizona	63	37	Nebraska	63	37
British Columbia	45	55	New Hampshire	63	37
California	61	39	New Jersey	78	22
Colorado	67	33	New Mexico	88	12
Connecticut	79	21	Nevada	41	59
District of Columbia	20	80	New York	75	25
Florida	60	40	Ohio	69	31
Georgia	71	29	Oklahoma	30	70
Hawaii	90	10	Ontario	20	80
Iowa	56	44	Oregon	58	42
Idaho	93	7	Pennsylvania	73	27
Illinois	78	22	Rhode Island	62	38
Indiana	61	39	South Carolina	63	37
Kansas	45	55	South Dakota	0	100
Kentucky	49.5	50.5	Tennessee	69	31
Louisiana	83	17	Texas	66	34
Massachusetts	60	40	Utah	80	20
Manitoba	60	40	Virginia	61	39
Maine	55	45	Washington	60	40
Michigan	78	22	Wisconsin	75	25
Minnesota	88	12	West Virginia	75	25
Missouri	67	33	Wyoming	70	30

In addition, an analysis was performed on the survey data to determine if the type of cross-connection control program being administered had any correlation to the percentages of indirect and direct cross-connections. Comparing internal cross-connection control programs (those aimed at preventing cross-connections at the point of use) with system cross-connection control programs (those aimed at controlling cross-connections at the service connection) and combination programs demonstrated that the type of program has no bearing on the types of cross-connections found in each system. The overall average remained 65% indirect cross-connections and 35% direct cross-connections with less than a 1% variance between the different programs being administered.

Additionally, a question in the survey was included to determine what kinds of cross-connections are most prevalent at the customers' point of use. Respondents were given several examples to choose from and the opportunity to indicate "other." Responding to the survey, 69% indicated finding cross-connections at irrigation systems: 51%, at garden or wash down hose uses; 48%, within fire suppression systems and 44% indicating cross-connections at boiler uses. Further analysis of the data show that 40% of the survey respondents indicated finding cross-connections at post mix carbonation units; 33%, at wells; 32%, at swimming pools; 29%, at food processing facilities; 27%, at cooling towers; 23%, at tank uses; 20%, at toilets and 10% indicated finding cross-connections at other unspecified points of use.

The utilities participating in the survey were asked to identify the three types of facilities at which they most commonly find cross-connections. Thirty-eight percent of the respondents included medical facilities; 37%, manufacturing; 37%, services (laundries, car washes, film labs, etc.); 36%, food processing; 36%, residential; 31%, schools and 7% included sewage treatment facilities. Additionally, the respondents, as a whole, indicated 55.5% of the cross-connections discovered were connected to non-health hazards; 39%, health hazards and 5.5%, sewage. They were also asked to determine the percentage of cross-connections found to be bacteriological or chemical in nature. Cross-connections that were bacteriological in nature accounted for 28.8% of the cross-connections; 26.1% were recorded as chemical in nature, and 29.8%, a combination of both. "Other" was indicated for 15.6% of the cross-connections.

HISTORY OF BACKFLOW INCIDENTS

The history of backflow incidents in North America was studied to help determine what has occurred. Correlating this information with other data obtained from the survey helped the research team determine those factors that may influence backflow incidents.

According to the survey results 38 % of the respondents to the initial survey indicated they did indeed experience backflow incidents while 42% indicated having not experienced an incident and 19% indicated "unknown." The specific number of incidents occurring since 1970 was requested with 51% of the respondents indicating one to two incidents since 1970; 23%, indicating 3-5; 10%, 6-10; 7%, 11-20; 4%, 21-50 and 2% indicated 51 to 100. Realizing that incidents are often not documented, the number of documented incidents was also requested with 66% indicating 1-2 documented incidents; 20%; 3-5, 8%, 6-10; six to ten; 3%, 11-20; 1%, 21-50; 0%, 51-100 and 1% indicating more than 100 documented incidents. The summary of this information is included in Table 3.3.

Table 3.3.
Relationship of number of incidents with response documentation

No. of Incidents	Percent Responding with Incidents	Percent of Documented Incidents
1-2	51%	66%
3-5	23%	20%
6-10	10%	8%
11-20	7%	3%
21-50	4%	1%
51-100	2%	0%
Over 100	4%	1%

To further enhance this data, the follow-up survey requested more detailed information, such as how many incidents had occurred within specific time frames. This information was correlated with the number of agencies having a cross-connection control program on paper and in practice. The data shows the number of agencies having a program in practice and those having a program on paper in 1975. Additionally, the number of agencies reporting backflow incidents from 1970 to 1975 and the average number of incidents being reported are shown. This analysis was performed for five-year periods up through the year 2000 and is summarized in Table 3.4.

Table 3.4.
History of Cross-Connection Control Programs since 1975

History				
Year	**Average No. of Incidents (in five years preceding)**	**No. of Utilities Reporting**	**No. of Utilities with Programs on Paper**	**No. of Utilities with Programs in Practice**
1975	4.3	9	23	20
1980	3.1	9	53	45
1985	2.2	14	80	77
1990	4.43	22	140	127
1995	3.13	30	198	179
2000	2.66	41	241	234

The data indicates that more and more agencies developed programs over the last three decades. As more agencies developed programs, more incidents were reported. Therefore, the reporting of incidents has continually increased since 1975. It is important, however, to note that the average number of incidents being reported per agency does not significantly change. This suggests that the number of incidents per utility isn't necessarily changing over time, but that reporting is becoming more prevalent as agencies develop cross-connection control programs.

Of the agencies with cross-connection control programs, 40% indicated they had had a backflow incident in the last 30 years, whereas 15% of the agencies without a program indicated they had had an incident. This does not imply that agencies with cross-connection control programs have more backflow incidents, but it further supports the data shown in Table 3.4, that agencies with cross-connection control programs report incidents more readily than agencies without a program. This may be due to the greater knowledge of the subject matter within the agencies with programs. Agencies without a program are less likely to be aware of backflow incidents and therefore, not as likely to report them.

The agencies reporting were asked to differentiate between those cross-connections, which took place on the premises of a customer, and those, which took place in areas directly under the control of the water distribution system. The agencies indicated that 83% of documented cross-connections were inside the premise of the customer, while 16% were in areas under the direct control of the water distribution system. When asked to determine the physical effects of the backflow incidents that occurred, 50% indicated none; 47% water quality complaints; 19%, interruption of service; 10%, public notice issued; 6%, replacement of distribution components; 5%, damage to the distribution system and 5% reported illnesses. These numbers total over 100%, since many of the agencies had more than one effect from the incidents that occurred.

FINANCIAL IMPACTS

Costs of Administering the Program

The estimated annual cost of administering a cross-connection control program was determined from the survey data along with the financial impact of backflow incidents. The results are broken down by utility size based on the number of service connections. The cost of the program is also broken down into personnel, equipment/material and "other." The overall average for all sizes of water utilities is $44,835 spent annually on administering the cross-connection control program. Personnel require $34,129 of this; equipment and material, $7,019 and $3,687 is spent on other items. The data shown in Table 3.5 indicates the average amount spent in each of these categories for various size utilities. On the average 3.65% of the agencies' operations and maintenance budget was directed towards cross-connection control.

Table 3.5
Annual cost of administering a Cross-Connection Control program

Annual Cost of Administering a Cross-Connection Control Program (U.S. Dollars)							
No. of Water Connections	**500**	**501-1000**	**1001-5000**	**5001-25,000**	**25,001-100,000**	**100,000**	**Overall**
Personnel	8,842	8,084	7,380	28,233	71,810	206,845	34,129
Equipment Material	4,122	2,949	1,939	6,181	15,949	42,526	7,019
Other	1,991	4,425	1,718	4,227	26,528	9,400	3,687
Total	14,995	15,458	11,037	38,641	114,287	258,771	44,835
Cost per service connection per year	At least $29.99	$15.46 to $30.85	$2.21 to $11.02	$1.55 to $7.72	$1.14 to $4.57	$2.59 or less	

Another aspect of the overall costs involved is the cost to purchase, install, test and repair a backflow prevention assembly. The utilities were asked to provide the average cost in their area of jurisdiction for these items. These costs are not necessarily incurred by the utility, but in most cases by the owner of the property where the assembly will be installed. This information is shown in Table 3.6.

Table 3.6.
Average costs for different backflow preventers

Assembly size	**1"**	**4"**	**10"**
Purchase	$159.57	$1241.92	$4839.06
Installation	$343.67	$1935.04	$5204.13
Cost of Testing	$56.78	$67.14	$69.41
Cost of Repair	$90.31	$207.55	$552.57

Costs of Backflow Incidents

The financial impact of backflow incidents was determined from the initial survey. The type of financial impact was broken down into several categories. Of the reporting agencies, 52% indicated no financial impact; 38%, investigative types of financial impact to the agency; 30%, analysis costs; 26%, clean-up costs; 3%, damage to the distribution system; 4%, a need to replace distribution components; 7%, lost revenues; 13%, having to purchase and/or install backflow preventers; 2%, other costs and 6% reported unknown.

Reporting agencies on the average indicated that, per incident, the cost to the agency was approximately $4,201 in direct costs, $1,292 in indirect costs and $10,650 in personnel costs for an overall average cost of $16,143 per incident.

The agencies also reported the estimated financial impact the backflow incidents had upon the consumer. No impact was reported by 39% of the agencies, 15% reported lost revenues; 5%, personnel costs; 23%, clean up costs; 41%, having to purchase and/or install a backflow preventer; 5%, other costs, and 11% reported unknown. The cost to the customer per incident, as estimated by the water utility, was $1,423 in direct costs, $1,750 in indirect costs and $600 in personnel, for an overall average of $3,773 in costs to the consumer per backflow incident.

It is important to note at this point that the range of expenses associated with a backflow incident was dramatic. Some agencies had figures as low as $25, $7, or $0, while some had expenses of over $100,000 with one even indicating $1.6 million.

The total number of backflow incidents since 1970 experienced by ~300 respondents (about ½ of the total respondents) range from a total of 2,800-4,100, with documented incidents ranging from 1,100-1,750. If extrapolated to the total number of public water systems in the United States (~170,000 – assuming ½ of them experienced incidents at the same rate as our respondents), there may have been as many as 800,000-1,100,000 backflow incidents experienced nationwide since 1970. Adding the $16,140 cost to the agency per incident as estimated above, the financial impact of all the incidents could be as high as thirteen to eighteen billion dollars. It is important to note that the figure noted above is not a statistically verifiable figure, due to the fact that only those choosing to respond to the survey did so. This disqualifies the data set as a true random sampling of the overall population pool. These figures are simply given as an example of the possible overall impact of cross-connections. Even with this in mind, the information is clear enough to indicate that cross-connections are a serious matter and should be addressed by water utilities.

HYDRAULIC CHANGES WITHIN A DISTRIBUTION SYSTEM

System Hydraulic Conditions

There are several conditions that may contribute to backflow within the distribution system. The surveys attempted to determine which conditions existed within systems and to what extent. Respondents were requested to indicate all of the conditions, which occur in their system. Therefore, the total percentage well exceeds 100%. Seventy-five percent of the respondents reported main breaks or leaks due to aging infrastructure, 62% indicated flushing program activity; 35%, main breaks or leaks due to human error; 33%, low or fluctuating line pressure; 29%, improper sized piping due to system growth; 27%, high demand due to fire fighting or other activities; 18%, indicated excessive elevation differences, and 15% indicated inadequate system storage. This information provides evidence that these types of hydraulic conditions are quite common in any distribution system.

Nominal working pressure in the distribution system was another condition about which the research team inquired. Five percent of the respondents reported a nominal working pressure of 20-40 pounds per square inch (PSI); 36%, 41-60 PSI; 36%, 61-80 PSI; 15%, 81-100 PSI; 5%, 101-125 PSI, and 3% indicated a nominal working pressure of greater than 125 PSI.

Information was gathered on elevation differences within the same distribution system. Respondents were asked what the maximum difference in elevation was in their distribution system. Eighteen percent indicated an elevation difference of less than 50 feet; 16%, 50 to 100 feet; 25%, 101 to 250 feet; 24%, 251 to 500 feet, and 17% indicated a maximum elevation difference of more than 500 feet.

Pressure fluctuation was another key hydraulic condition, which was analyzed. The respondents were asked what the typical pressure fluctuation was in their system throughout the day. Sixteen percent indicated a pressure fluctuation of less than 5 PSI; 43%, 5 to 10 PSI; 31%, 11 to 20 PSI; 5%, 21 to 30 PSI; 2%, 31 to 40 PSI; and 2% indicated a typical pressure fluctuation of more than 40 PSI throughout the day. One piece of data, which relates directly to backflow, was the reporting of water meters running backwards. Seventy-seven percent of the respondents indicated they had a means of determining if their water meters run backwards. Those that indicated they had the means to determine if the meters run backwards also indicated an average of 5.2 meters documented as running backwards in the prior twelve months. The average number of meters documented as running backwards and indicating the potential for backflow was also recorded for 1995, 1996, 1997, and 1998. The average number for each of these years was 2.5, 2.2, 2.2, and 6.0 respectively. Eighty-five percent indicated the information was determined from billing information, while 15% indicated other methods, such as inspection reports, service orders, meter reading, and distribution employees.

Hydraulic Conditions as Related to Backflow Incidents

The research team analyzed various data from the survey in an attempt to determine which hydraulic conditions reported by the responding utilities had an effect on the number of backflow incidents reported by the utilities. After correlating several water system parameters with the number of backflow incidents reported, some important facts became evident. In the analysis, the research team determined if certain hydraulic conditions were more evident than others in those systems, which reported more than ten backflow incidents in the same time frame (since 1970).

The data analysis showed that the nominal working pressure in a water system does have an effect on the number of backflow incidents experienced. Systems with a nominal working water pressure of more than 100 PSI, experienced less backflow incidents than those systems with lower nominal water pressures. For example, the majority of respondents indicated their system working pressure lies between 40 and 100 PSI. In these systems, between 13 and 20 percent of the respondents experienced over ten backflow incidents since 1970 and 18% of the respondents with less than 40 PSI nominal working pressure experienced more than ten incidents since 1970. Of respondents recording nominal working water pressures of 100 PSI or greater, just over 6% experienced over ten backflow incidents in the same time period.

Also analyzed was the nominal working pressure at the most remote point in the distribution system. Differences in this pressure had no significant effect on the number of backflow incidents experienced.

The data analysis demonstrated that those utilities with higher maximum elevation differences experienced a greater number of backflow incidents. The analysis showed that 6% of water utilities with a maximum elevation difference of 50 feet or less in the distribution system experienced more than ten backflow incidents since 1970. Of the utilities with a maximum elevation difference of between 100 and 250 feet, 14% indicated they had experienced more than ten backflow incidents in the same time frame. Finally 27% of the utilities with a maximum elevation difference of 500 feet or more indicated experiencing more than ten backflow incidents since 1970.

Similarly, another definite trend was discovered involving the pressure fluctuations experienced within a distribution system. Of the utilities indicating a pressure fluctuation of between five and ten PSI, 11% indicated experiencing more than ten backflow incidents during the study time frame. Of the utilities with more than 40 PSI of pressure fluctuation, 50% indicated having experienced more than ten backflow incidents. Analyzing a combination of nominal system pressure and maximum elevation difference indicated that systems with lower nominal pressure and higher elevation differences experienced more incidents than systems with higher nominal pressure and lower elevation differences. Six percent of those systems with a nominal working pressure of 20 to 60 PSI reported more than ten backflow incidents in the study time frame when the maximum elevation difference in the system was 250 feet or less. With the same system pressure, but the maximum elevation difference above 250 feet, 12% reported more than ten incidents. With the nominal system pressure at 60 to 125 PSI, 3% of those utilities with maximum elevation differences of less than 250 feet indicated more than ten incidents, whereas 11% with the same nominal system pressure indicated more than ten incidents when the maximum elevation difference was greater than 250 feet.

As expected, the larger water utilities experienced more backflow incidents. Of the water utilities, which had backflow incidents and 500 service connections or less, only 5% indicated having more than ten incidents in the study time frame. Of those with 100,000 or more service connections, 56% of those reporting backflow incidents reported ten or more. The chart in Table 3.7 provides the details.

Table 3.7.
Reported incidents as a function of number of service connections

No. of Service Connections	% of Utilities Reporting more than Ten Incidents (of those reporting incidents)
<500	5
501-1000	0
1001-5000	10
5001-25,000	15
25,001-100,000	25
>100,000	56

CROSS-CONNECTION CONTROL PROGRAMS

Overview of Cross-Connection Control Programs

The research team gathered various data in an attempt to determine what is included in an effective cross-connection control program.

Ninety-one percent of the utilities responding to the survey indicated that they did have a cross-connection control program. The respondents were asked to identify which sources or recommended practices they use. Fifty-five percent indicated they use the USC *Manual of*

Cross-Connection Control, Ninth Edition; 49%, AWWA's *Manual M-14;* 27%, the *EPA Cross-Connection Control Manual,* and 24% indicated, "other." The numbers demonstrate that utilities follow more than one source or recommended practice.

Eighty percent of the utilities indicated they conduct site surveys to determine the extent of cross-connections on a premise as a part of their cross-connection control program. Of those that indicated they did not conduct site surveys, 78% indicated they had a blanket program, requiring backflow prevention on specific types of services without a survey. Another administrative authority was responsible for carrying out the site survey in 5.6% of the reporting utilities. A third party was responsible for site surveys in 3.4% of the cases, , 2.4% indicated a survey-by-mail was used, and 1% conducted phone surveys.

The survey data shows that 43% of the utilities surveyed administered system protection programs, meaning their goal is to protect the water system at the users water connection. Only 6% indicated that they had an internal protection program, where the water systems are protected at the point of water use. A combination of these methods was indicated by 47% of those responding to the survey.

Those agencies which used system protection indicated the following number (on average) of backflow preventers in their system: 7.5 air gaps, 446 reduced pressure principle assemblies, 17 reduced pressure principle detector assemblies, 285 double check valve assemblies, 110 double check detector assemblies, 48 pressure vacuum breaker assemblies, 7 atmospheric vacuum breakers and 18 "others." The follow-up survey sought to clarify "others." When asked to specify what this meant, the respondents generally responded with: dual checks, hose bib vacuum breakers, and single check valves.

Agencies administering internal cross-connection control programs indicated the following number (on average) of backflow preventers in their system: 33 air gaps, 130 reduced pressure principle assemblies, 129 reduced pressure principle detector assemblies, 187 double check valve assemblies, 5 double check detector assemblies, 44 pressure vacuum breakers, 1 atmospheric vacuum breaker, and 30 "other." For a combination program, one where the utility administers both an internal program and a system program, the following assemblies were listed on the average: 1139 air gaps, 437 reduced pressure principle assemblies, 6 reduced pressure principle detector assemblies, 290 double check valve assemblies, 75 double check detector assemblies, 195 pressure vacuum breaker assemblies, and 43 atmospheric vacuum breakers.

The survey asked which elements of a cross-connection control program were included in the utilities' program. The choices were as follows:

- Legal authority
- Acceptable backflow preventers
- A requirement to field test backflow preventers
- An acceptable method of field-testing backflow prevention assemblies
- A means to determine who is qualified to field test backflow prevention assemblies
- Conducting site surveys to determine the extent of cross-connections on a premise
- A means to determine who is qualified to conduct site surveys
- A requirement for reporting backflow incidents
- A public education effort

A legal authority was included in 93% of the cross-connection control programs indicating this was the most included element in cross-connection control programs. The next most included element was an acceptable list of backflow prevention assemblies, with 90% of the utilities responding to the survey indicating that they did have a requirement for acceptable backflow prevention assemblies. Of those responding, 46% required assemblies approved by the University of Southern California (USC); 12%, required assemblies listed by the American Society of Sanitary Engineering (ASSE); 9%, Underwriters Laboratories (UL); 7%, the International Association of Plumbing and Mechanical Officials (IAPMO); 4%, Factory Mutual (FM): 3%, the Canadian Standards Association (CSA), and 19% indicated "other."

A requirement to field test backflow preventers was indicated by 89% of the utilities responding, indicating that almost all utilities requiring a specific approval or listing of backflow prevention assemblies also has a requirement to field test the assemblies. The field test procedures published in either the Eighth or Ninth Edition of the USC *Manual of Cross-Connection Control* was used by 74% of those responding to the survey. The ASSE Series 5000 was used by 9%; the New England Water Works Association method was used by 8%, and 19% indicated "other."

A requirement to keep records of the backflow preventer field tests was indicated by 84%. Although not an incredibly large difference, it is significant to note there are a number of utilities that require backflow preventers to be tested, but don't require records to be made or kept regarding the field test results. Again there is a discrepancy in the number of utilities requiring field tests of backflow preventers and those requiring some means of determining who is qualified to conduct such field tests (most commonly found in the form of a certification program). Although 89% of the utilities indicated that there was a requirement to field-test backflow preventers, only 81% indicated that there was a means to determine who was qualified to perform such tests, and 70% had a requirement for a specific acceptable method for testing the backflow preventers.

Site surveys were required by 80% of the utilities to determine the extent of cross-connections on a user's property. However, only 57% had a means to determine who was qualified to conduct such site surveys.

When a backflow incident does occur, the survey indicates that only 49% of the utilities require that the incident be reported. Additionally, only 42% have a public education effort as part of their program. A summary of these findings is shown in Table 3.8.

Table 3.8
Inclusion of Various Elements in Cross-Connection Programs

Ranking	% Reporting	Element
1	93	Legal Authority
2	90	List of Acceptable Backflow Preventers
3	89	A Requirement to Field-test Backflow Preventers
4	84	A Requirement to Keep Records of Backflow Preventer Field Test Results
5	81	A Means to Determine who is Qualified to Test Backflow Preventers
6	80	Conducting Site Surveys
7	70	An Acceptable Method of Field-testing Backflow Preventers
8	57	A Means to Determine Who is Qualified to Conduct Site Surveys
9	49	A Requirement for Reporting Backflow Incidents
10	42	A Public Education Effort to Educate the Public in the Area of Cross-Connection Control

EFFECTIVENESS OF THE ELEMENTS OF A CROSS-CONNECTION CONTROL PROGRAM

The respondents were asked to rank the effectiveness of the various elements of a cross-connection control program. A ranking of "1" indicated the most effective. The average ranking of all the respondents is shown in the chart of Table 3.9.

Table 3.9.
Relative effectiveness of various elements in Cross-Connection Control Programs

Ranking	Element in Program
1.6	Legal Authority
3.0	List of Acceptable Backflow Preventers
3.4	A Requirement to Test Assemblies
3.7	Conducting Site Surveys
4.2	A Means to Determine who is Qualified to Test Backflow Preventers
4.6	An Acceptable Method of Field-testing Backflow Preventers
5.1	A Requirement to Keep Records of Backflow Preventer Field Test Results
5.6	A Means to Determine Who is Qualified to Conduct Site Surveys
6.3	A Public Education Effort to Educate the Public in the Area of Cross-Connection Control
6.7	A Requirement for Reporting Backflow Incidents

REGULATIONS AND ENFORCEMENT ISSUES

Ninety-three of the respondents indicated that they had legal authority to carry out their cross-connection control program, and "Legal Authority" was listed as the most effective element of the cross-connection control programs. Information was gleaned from the survey results in order to determine exactly what legal authority meant. The respondents were asked what elements were contained in their legal document. The most common item, noted by 91% of the respondents, was the requirement to field-test backflow preventers periodically. The survey indicated that 6% of the respondents require field-testing of backflow preventers every six months. Eighty-three percent indicated backflow preventers must be tested every twelve months. Three percent required testing every twenty-four months. Four percent indicated "other" and 8% indicated there was no requirement. It should be noted that several of those that marked "other," wrote in, "annually." Eighty percent indicated they had a requirement for some type of listing or approval for backflow preventers. Certification or licensing of backflow prevention assembly testers was required in 78% of the legal documents. Likewise, 78% of the respondents' legal documents required a site survey to determine if backflow protection was necessary. Also required by 78% of the legal documents was to make and keep, for a period of time, records of backflow assembly installations, field tests, and repairs. Fifty percent had a specific method for field-testing backflow preventers spelled out in their legal document. Only 43% had a requirement for reporting backflow incidents.

According to the survey data, respondents utilize the following in bringing customers into compliance with their backflow prevention program. Eighty-nine percent use compliance letters and written notices; 57%, termination of service; 33%, public education; 12%, fines; 5%, civil citations; 2%, criminal citations, and 4% responded "other." The next correlation was how many respondents utilize compliance letters and written notices and what those respondents felt was most effective. Ninety-one percent utilize compliance letters. However, only 55% consider them the most effective means of gaining compliance. Furthermore, analysis looked at those who utilize termination of service or threat thereof as an effective means of gaining compliance. Fifty-three percent of the respondents indicate using termination of service as a means of gaining compliance, while only 44% indicate this as the most effective means of gaining said compliance.

Next survey respondents were asked to rank what sort of enforcement actions they consider to be most effective. As stated above, 55% indicated compliance letters and written notices. Forty-five percent indicated termination of water service. Only 15% indicated public education and 5% indicated fines. None considered criminal citations to be effective and only 1% considered civil citations effective. Seven percent responded "other."

PUBLIC EDUCATION

A public relations effort or program can consist of many things. Public education, public relations, investor relations, crisis management, general communications, customer service, can all play a part in a public relations effort. The survey did assess many aspects of an overall public relations effort; however, public education was specifically touched upon. Respondents indicated that 60% of their customers believe that a cross-connection control program is not needed, specifically indicating that there is no perceived risk. However, only 44% of those responding have ever tried some type of public education program, while only 14% of respondents believe public education is the most effective way to bring a customer into compliance.

It is believed, by many involved in cross-connection control, that public education is the most effective means of bringing customers into compliance. This survey helps determine the accuracy of this statement.

According to the survey data, only 42% of the respondents have a public education effort in place as part of their cross-connection control program. That is one of the least included elements of all possible elements included in the survey questionnaire. When asked about the effectiveness of a public education effort, it was ranked second to last in the list of ten elements included in the survey. Only 15% of the survey respondents ranked public education as the most effective method of enforcement. This data can be compared with responses from another survey question. When asked what kind of complaints are encountered by agencies in administering their cross-connection programs, 60% of the respondents indicated that the #2 customer complaint is no perceived risk by the customer—they feel the program is unneeded. This is second only to cost, with 82% of the respondents indicating customer cost complaints. Furthermore, when asked what kinds of limiting factors agencies come across in implementing or running a program, the second most experienced problem is uninformed customers with 37% of the total respondents indicating this.

While 60% of the respondents indicate their customers having no perceived risk or need of a program, only 44% of those have ever tried public education and only 14% believe public education to be the most effective means of gaining customer compliance.

CUSTOMER COMPLAINTS

An analysis of customer complaints was made to determine if there is correlation between the complaints and backflow incidents. The complaints were also analyzed to determine customer resistance to the administration of the program.

According to the survey results we found that 82% of the survey respondents received a customer complaint about the cost experienced as a result of the agency conducting or administering a cross-connection control program. Sixty percent of the survey respondents received a customer complaint about the program being unneeded and that there was no perceived risk of cross-connections. Forty-one percent of the survey respondents received complaints about a reduction in water pressure due to backflow preventers. Thirty-five percent of the survey respondents indicated that customers complained that cross-connection control is governmental intervention while 27% indicated receiving complaints that customers noticed a reduction in available water due to backflow prevention devices. This likely is also an indication of reduced pressure as opposed to a reduction in the quantity of water available. Thirteen percent of the respondents complained about an interruption of their water service due to periodic testing and 8% indicated other unspecified complaints.

CHAPTER 4
COMPUTER MODELING

GOVERNING EQUATIONS FOR COMPUTER SIMULATION

The computer simulation of the piping network for the present study contains the steady state analysis as well as transient analysis. A steady state simulation is first achieved, and then transient state simulation (caused by a sudden change in water demand or sudden change in hydraulic head at certain locations) is performed. After the transient state, a new steady state is reached.

Fluid mechanics principles and the governing equations used for the computer simulation will be briefly outlined below:

Governing equations for "steady state" piping network analysis are the continuity equation as well as the energy equation.

(1) At every pipe junction, the continuity condition must be satisfied. Flows into a junction must be equal to the sum of all flows out of a junction.

$$Q_{in} = \sum Q_1 + Q_2 + \cdots\cdots Q_j \tag{1}$$

Where Q_{in} is the flow rate into a junction, and $Q_1, Q_2 \cdots\cdots Q_j$ are flow rates in pipes out of the junction and demands at the junction.

(2) In any given closed loop, the sum of all energy losses must equal to zero.

$$\sum_{i=1}^{N} (H_L)_i = 0 \tag{2}$$

where $(H_L)_i$ represents the energy loss in the ith pipe in the closed loop and N represents the number of pipe segments contained in the closed loop.

These two sets of equations provide enough equations to solve for all the unknowns in the simulation of steady state piping network in steady state condition.

For the transient condition, when changes of flow are rapid, the fluid compressibility has to be taken into consideration. As a result, the flow changes are not experienced instantaneously throughout the system. The pressure waves move back and forth in the piping system. Using the momentum theorem, the change in hydraulic head can be related to the change in velocity by the following equation:

$$\Delta H = \frac{c}{g} \Delta V \tag{3}$$

where c : wave speed
g : gravitational acceleration
V : flow velocity
H : hydraulic head

For the transient analysis of a piping network, we need to satisfy all the equations describing these flows. Certain assumptions are needed for the analysis: The fluid is assumed slightly compressible, the walls of the conduit are linearly elastic and are slightly deformable, and the head losses during the transient state may be computed using the Darcy-Weisbach equation in terms of the instantaneous velocity in the pipe. These assumptions have been well adopted for all transient state analysis programs in the field.

The governing continuity equation for transient flow analysis is:

$$\frac{\partial P}{\partial t} + V\frac{\partial P}{\partial x} + \rho c^2 \frac{\partial V}{\partial x} = 0 \tag{4}$$

where P : pressure
V : average velocity in the pipe
ρ : fluid density
c : wave speed
x : spatial variable
t : time variable

The governing momentum equation for transient flow analysis is:

$$\frac{\partial V}{\partial t} + V\frac{\partial V}{\partial x} + \frac{1}{\rho}\frac{\partial P}{\partial x} + f\frac{V|V|}{2D} = 0 \tag{5}$$

where V : flow velocity
D : pipe diameter
P : pressure
ρ : fluid density
f : Darcy-Weisbach friction factor
$|V|$: the absolute value of the velocity

Equation (4) and (5) can be further simplified by eliminating the non-linear terms $V\frac{\partial V}{\partial x}$, and $V\frac{\partial P}{\partial x}$ because these terms are very small as compared to other terms. By substituting the pressure term with the hydraulic head ($P = \rho g H$), we obtain the simplified governing equations for transient flow analysis as follows:

$$\frac{\partial H}{\partial t} + \frac{c^2}{gA}\frac{\partial Q}{\partial x} = 0 \tag{6}$$

$$\frac{\partial Q}{\partial t} + gA\frac{\partial H}{\partial x} + \frac{f}{2DA}Q|Q| = 0 \tag{7}$$

where H : hydraulic head
Q : flow rate
c : wave speed
A : cross-sectional area
D : diameter of the pipe
g : gravitational acceleration
$|Q|$: the absolute value of the flow rate

RESULTS OF COMPUTER SIMULATIONS

Model City C (Small Network System)

As part of the project, the unsteady piping network analysis is performed for three different sized networks, representing three piping networks for three model cities.

For the smallest network (named the Model City C), a simple pipe network was used. A sketch of the pipe network is shown in Figure 4.1. This simple network has ten pipes, seven junctions and two reservoirs. This network is the same as the example problem in Watters' book (1984). We used this textbook example in order to verify some of the computations in the early stage of the study. The demands at Node #2, #3, #4, #6, #7 are 3cfs, 4cfs, 1cfs, 0cfs and 2cfs respectively.

Table 4.1 provides the junction report showing the demand and hydraulic grades at the seven junctions. Table 4.2 shows the steady state solution of the computed discharge in each of the ten pipes.

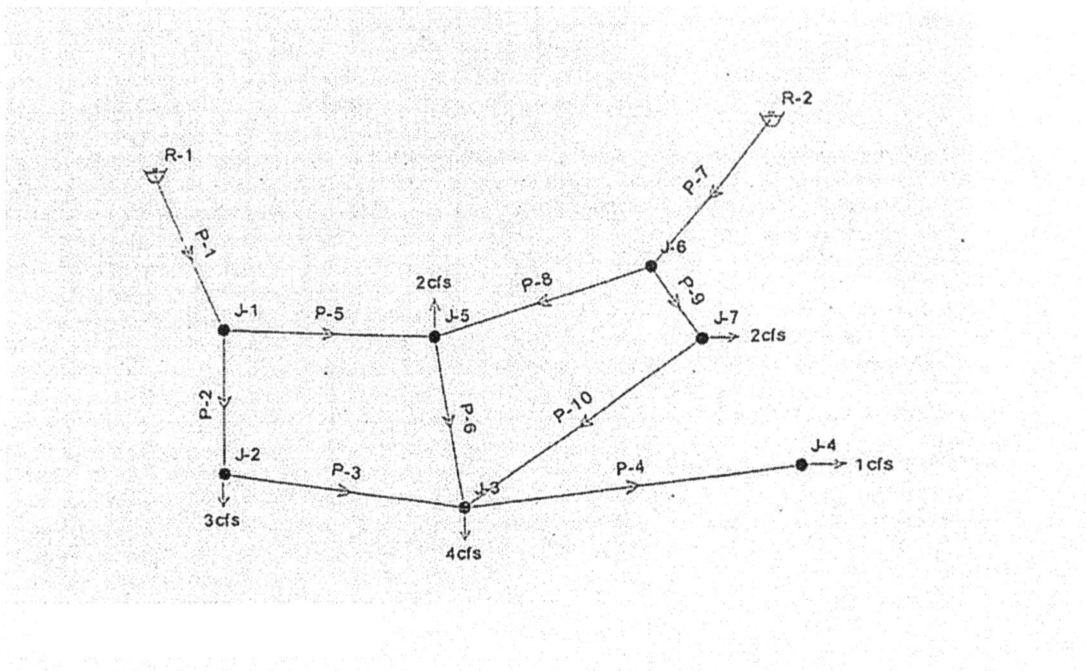

Figure 4.1 Pipe network

Table 4.1
Junction Report

Label	Elevation (ft)	Demand (ft^3/s)	Pattern	Calculated Demand (ft^3/s)	Calculated Hydraulic Grade (ft)	Pressure (psi)
J-1	1300.00	0.00000	Fixed	0.00000	1470.34	73.66
J-2	1290.00	3.00000	Fixed	3.00000	1461.84	74.31
J-3	1280.00	4.00000	Fixed	4.00000	1458.42	77.15
J-4	1260.00	1.00000	Fixed	1.00000	1450.72	82.47
J-5	1310.00	2.00000	Fixed	2.00000	1464.14	66.65
J-6	1320.00	0.00000	Fixed	0.00000	1467.62	63.83
J-7	1310.00	2.00000	Fixed	2.00000	1459.00	64.43

Table 4.2
Pipe Report

Label	Length (ft)	Diameter (in)	Material	Discharge (ft^3/s)	Upstream Structure Hydraulic Grade (ft)	Downstream Structure Hydraulic Grade (ft)	Pressure Pipe Headloss (ft)	Headloss Gradient (ft/1000ft)
P-1	800.00	15.0	Cast iron	8.89657	1480.00	1470.34	9.66	12.07
P-2	600.00	12.0	Cast iron	5.36928	1470.34	1461.84	8.50	14.17
P-3	1200.00	12.0	Cast iron	2.36928	1461.84	1458.42	3.42	2.85
P-4	1800.00	8.0	Cast iron	1.00000	1458.42	1450.72	7.69	4.27
P-5	1000.00	12.0	Cast iron	3.52729	1470.34	1464.14	6.20	6.20
P-6	800.00	10.0	Cast iron	2.34607	1464.14	1458.42	5.72	7.15
P-7	1000.00	10.0	Cast iron	3.10343	1480.00	1467.42	12.38	12.38
P-8	1200.00	8.0	Cast iron	0.81877	1467.62	1464.14	3.48	2.90
P-9	400.00	8.0	Cast iron	2.28465	1467.62	1459.00	8.62	21.55
P-10	1500.00	8.0	Cast iron	0.28465	1459.00	1458.42	0.58	0.39

Figure 4.2 shows the pipe network modified by the change in demand at Node #3 and Node #5. Demand at Node #3 is linearly decreased from 4cfs to 0cfs within 4 seconds; simultaneously, demand at Node #5 is linearly increased from 2cfs to 6cfs in this same 4-second period. (1cfs=448.43gpm)

The junction and pipe report with the demand change at Node #3 and Node #5 are presented in Table 4.3 and 4.4. Due to these changes of demand, the flow direction at Pipe #6 and Pipe #10 are changed into the opposite direction as shown in the time dependent discharge curve in Figure 4.3. From Figure 4.3, it is seen that the flow direction is reversed in Pipe #6 after 3.2 seconds, before the complete implementation of the change in demand at Node #3 and #5.

Figures 4.4 and 4.5 show the transient pressure head fluctuation at Node #5, #3, and #7 as a function of time after the implementation of the change in demand at Node #5 and Node #3 is accomplished in 4 seconds. It is noted that the pressure head at Node #3 is significantly increased and that the pressure head at Node #5 is significantly decreased with many cycles of oscillations. In order to detect the effect of different valve open/closure rate on the transient pressure fluctuation, computer simulation is also performed for the network shown in Figure 4.2 with the changing demands accomplished in 2 seconds. The corresponding results are shown in Figures 4.6, 4.7, and 4.8 (correspond to Figures 4.3, 4.4, and 4.5 for 4 second open/closure time).

By comparing Figures 4.3, 4.4, and 4.5 with Figures 4.6, 4.7, and 4.8, it is very clear that the rapid change of the demands at Node #3 and Node #5 produces a greater fluctuations of the flow rate and pressure heads.

In Figures 4.9 and 4.10, the effect of 2-second open/closure time is further compared with 4-second open/closure time. In each of these two figures, the 2-second open/closure time causes greater magnitude of the transient fluctuation. The range of peak to trough is significantly increased when the change of demand is accomplished in 2 seconds as compared to that in 4 seconds.

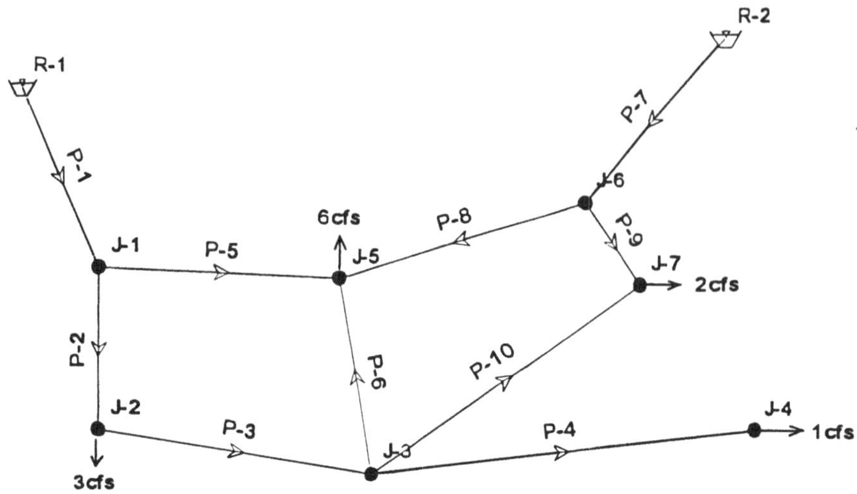

Figure 4.2 Pipe network with demand changes at Node #3 and Node #5

Table 4.3
Junction report with demand changes at Node #3 and Node #5

Label	Elevation (ft)	Demand (ft^3/s)	Pattern	Calculated Demand (ft^3/s)	Calculated Hydraulic Grade (ft)	Pressure (psi)
J-1	1300.00	0.00000	Fixed	0.00000	1470.25	73.62
J-2	1290.00	3.00000	Fixed	3.00000	1463.67	75.10
J-3	1280.00	0.00000	Fixed	4.00000	1461.85	78.64
J-4	1260.00	1.00000	Fixed	1.00000	1454.16	83.96
J-5	1310.00	6.00000	Fixed	2.00000	1461.39	65.47
J-6	1320.00	0.00000	Fixed	0.00000	1467.96	63.98
J-7	1310.00	2.00000	Fixed	2.00000	1461.80	65.64

Table 4.4
Pipe report with demand changes at Node #3 and Node #5

Label	Length (ft)	Diameter (in)	Material	Discharge (ft^3/s)	Upstream Structure Hydraulic Grade (ft)	Downstream Structure Hydraulic Grade (ft)	Pressure Pipe Headloss (ft)	Headloss Gradient (ft/1000ft)
P-1	800.00	15.0	Cast iron	8.94006	1480.00	1470.25	9.75	12.19
P-2	600.00	12.0	Cast iron	4.71183	1470.25	1463.67	6.57	10.96
P-3	1200.00	12.0	Cast iron	1.71183	1463.67	1461.85	1.82	1.52
P-4	1800.00	8.0	Cast iron	1.00000	1461.85	1454.16	7.69	4.27
P-5	1000.00	12.0	Cast iron	4.22823	1470.25	1461.39	8.85	8.85
P-6	800.00	10.0	Cast iron	-0.63675	1461.39	1461.85	0.46	0.58
P-7	1000.00	10.0	Cast iron	3.05994	1480.00	1467.96	12.04	12.04
P-8	1200.00	8.0	Cast iron	1.13502	1467.96	1461.39	6.56	5.47
P-9	400.00	8.0	Cast iron	1.92492	1467.96	1461.80	6.15	15.38
P-10	1500.00	8.0	Cast iron	-0.07508	1461.80	1461.85	0.05	0.03

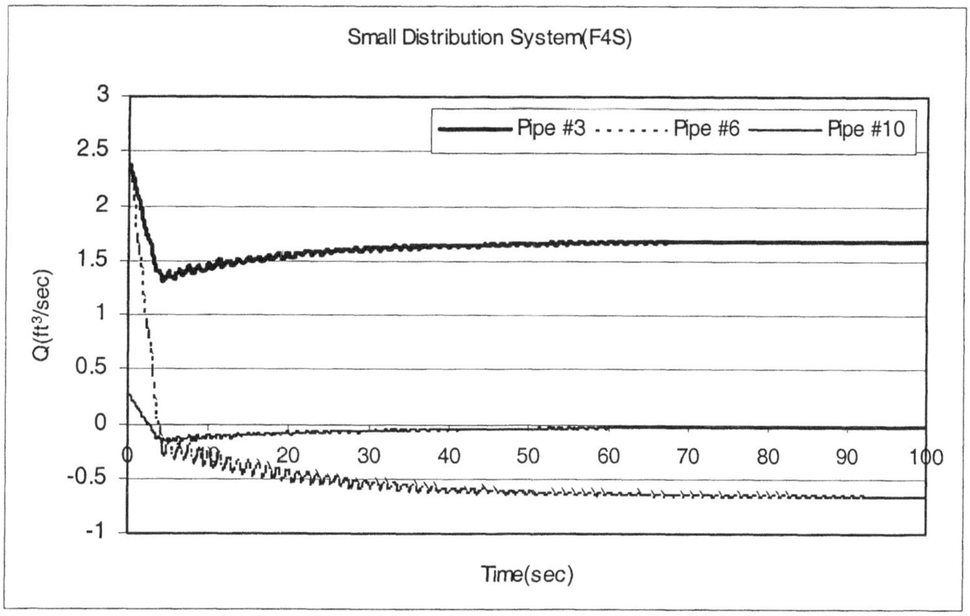

Figure 4.3 Transient flow rates at Pipe #3, #6, #10 when the change of demand is implemented within 4 seconds

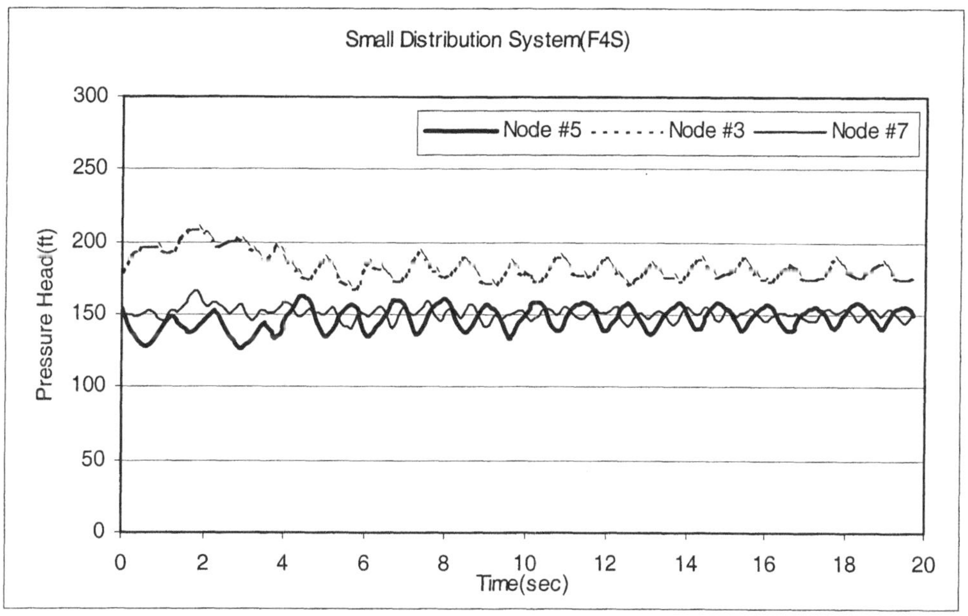

Figure 4.4 Transient pressure head fluctuations at Node #5, #3, #7 after the change in demand is implemented in 4 seconds

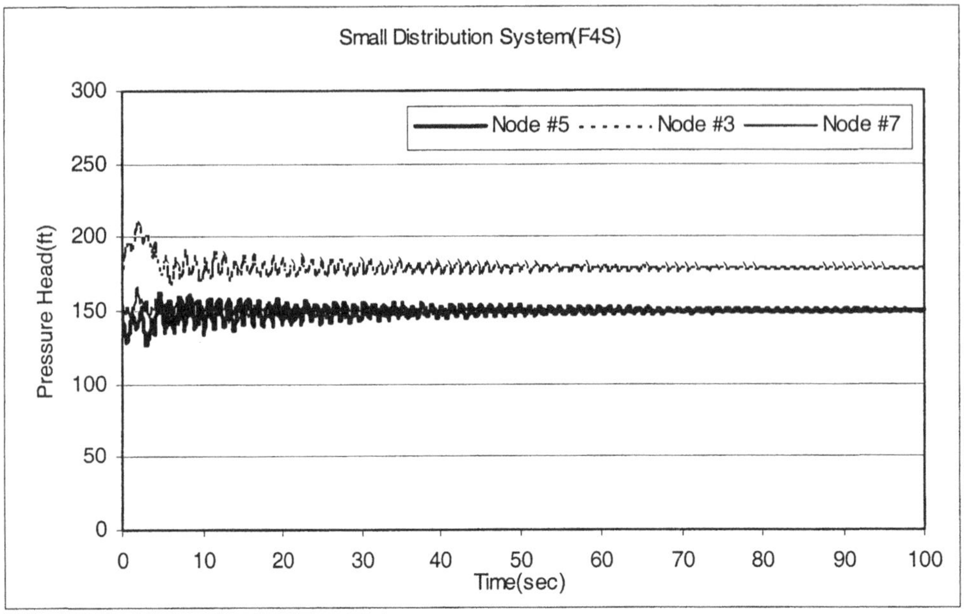

Figure 4.5 Transient pressure head fluctuations at Node #5, #3, #7 for the first 100 seconds after the change in demand is implemented in 4 second

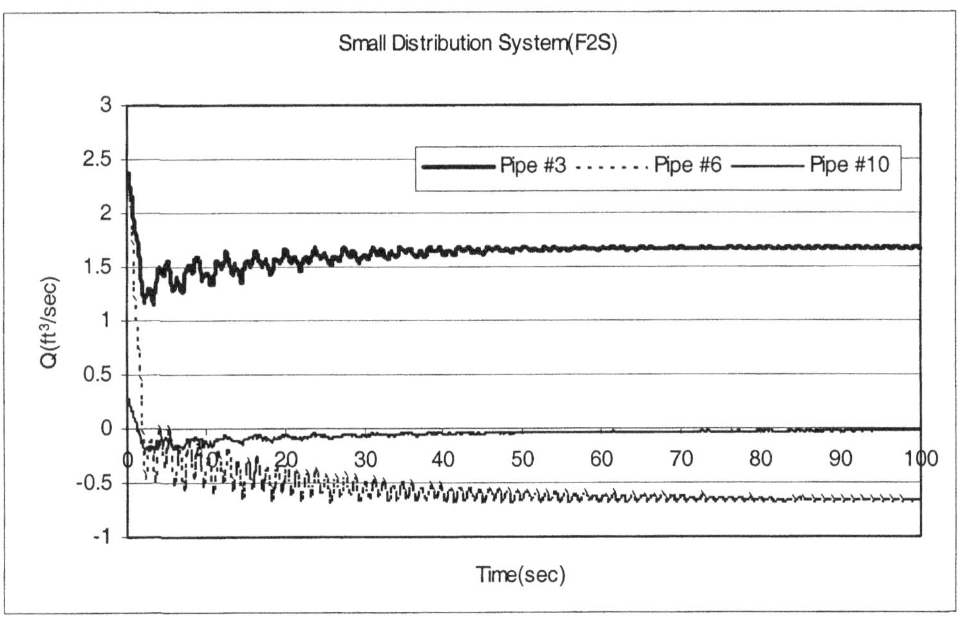

Figure 4.6 Transient flow rates at Pipe #3, #6, #10 when the change of demand is implemented within 2 seconds

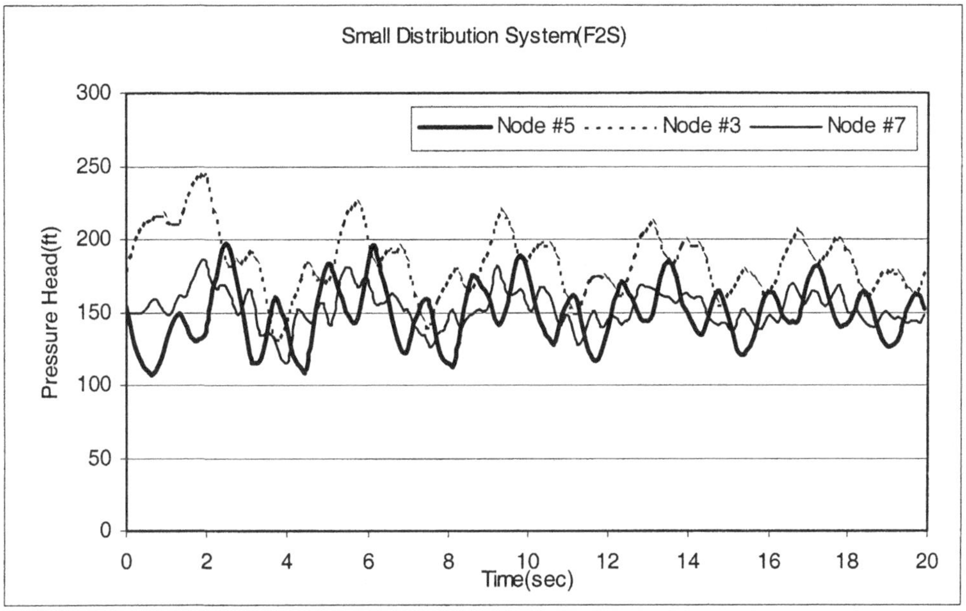

Figure 4.7 Transient pressure head fluctuations at Node #5, #3, #7 after the change in demand implemented in 2 seconds

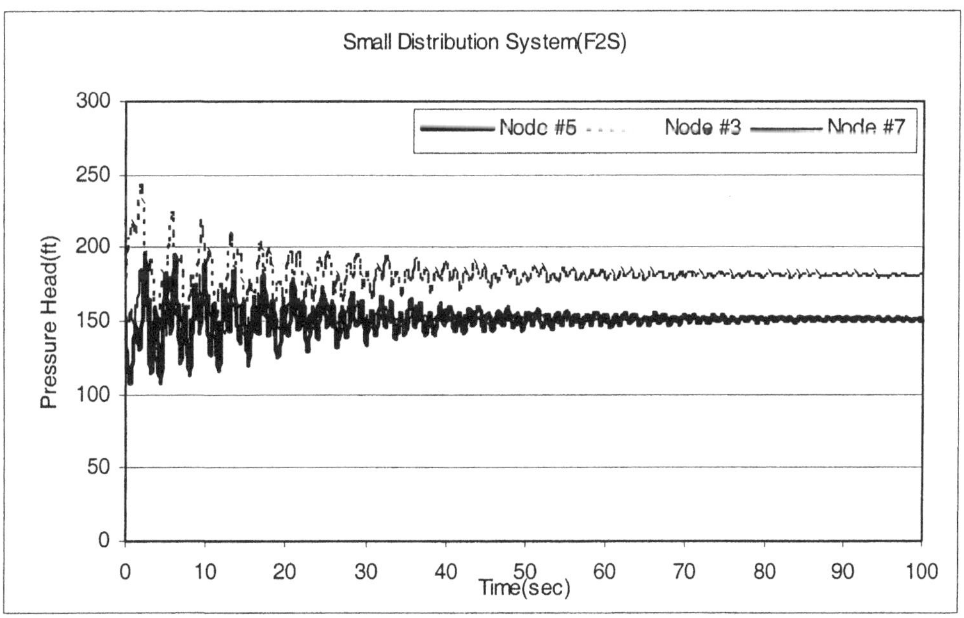

Figure 4.8 Transient pressure head fluctuation at Node #5, #3, #7 for the first 100 seconds after the change in demand is implemented in 2 seconds

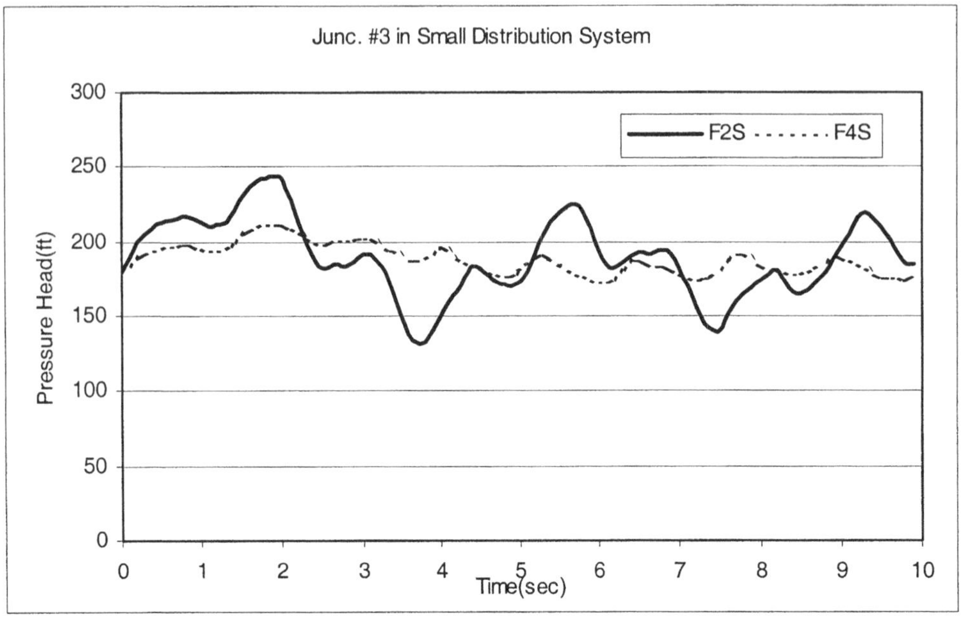

Figure 4.9 Comparison of pressure head fluctuations at Node #3 showing the effect of different valve open/closure times (2 sec vs. 4 sec)

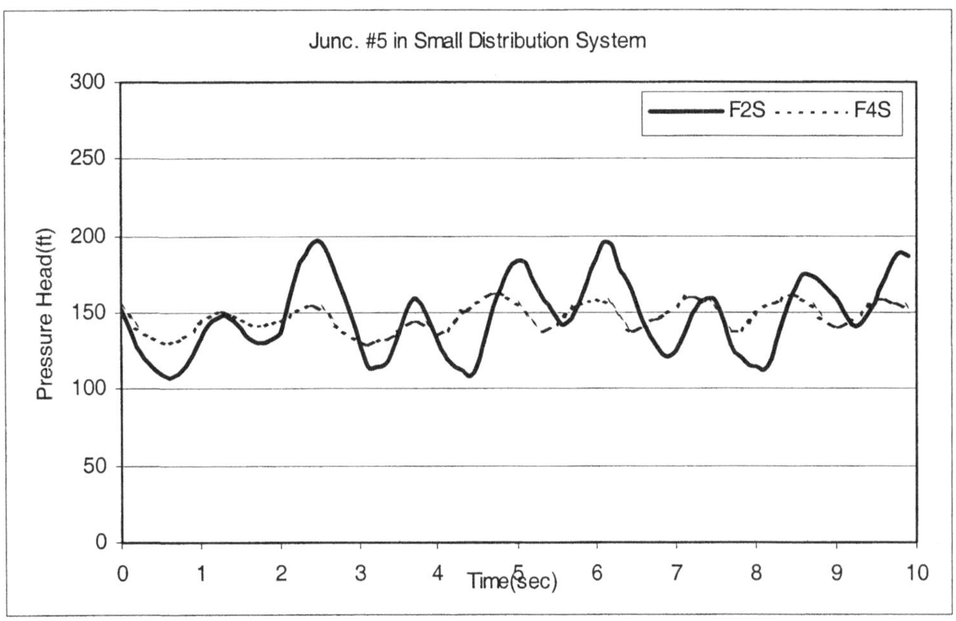

Figure 4.10 Comparison of pressure head fluctuations at Node #5 showing the effect of different valve open/closure times (2 sec vs. 4 sec)

Model City B (Medium Sized Network System)

To simulate a medium sized system, an existing water distribution system in the mid-west region of the United States is used. The piping network supply 36cfs of flow by two pump stations and a reservoir. Most of the nodes are demands that take flow from the system. The system consists of 252 nodes and 371 pipes. Three different simulation conditions are conducted. The system characteristics at the nodes and the pipeline for Simulation No. 1 for the steady state condition (before the start of the transient analysis) are presented in Appendix 5. Also included in this appendix is the connectivity of the nodes and the pipes used for the simulation.

Simulation No. 1 (Model City B)

For the transient analysis, it is assumed that a hydrant at Node #201 opened in two seconds creating a sudden demand of 1121gpm at this location. Figure 4.11 and 4.12 show the schematics of the part of the system in the immediate vicinity of Node #201 before and after the hydrant at Node #201 opens.

The transient flows generated by the opening of the hydrant at Node # 201 are quite dramatic as shown in Figure 4.13. The figure shows the pressure head time history at three nodes: Node #125, Node #200, and Node #10. The location of Node #200 is approximately 1200ft from the hydrant. In reviewing Figure 4.13, it is seen that the pressure head at Node #200 initially drops 80ft (from 140ft to 60ft) and subsequently drops down to 20ft level. At Node #125 (which is located near the upper right corner in Figure 4.11), the magnitude of the pressure drop is smaller but still quite dramatic, dropping from around 180ft to 120ft before it recovers gradually. At Node #10 (located near the lower left corner shown in Figure 4.11) which is fairly well isolated from the hydrant location, a gradually pressure drop of approximately 20ft is experienced. From Figure 4.13, it is seen that the pressure changes are more immediate and the most dramatic for locations closer to the location of the opening hydrant. At a location further away, the transient conditions are felt at a later time and are less dramatic. These results clearly indicate that a sudden demand change, such as the opening of a fire hydrant, can have significant effect on the flow and pressure in a distribution system. Cross-connection control becomes even more important for safe guarding the potable water in the distribution system whenever such sudden demand change may occur. A sudden pressure drop of such magnitude can cause backflows and if a cross-connection exists it can certainly create undesirable consequences, compromising the integrity of the potable water system. A similar result of pressure at Node #6, Node #205, and Node #203 is presented in Figure 4.14. The pressure drops at Node #205 and Node #203 are quite drastic as they are located closer to the open hydrant location. The pressure drop at Node #6 (which is located further away from Node #201) is noticeable but not quite as drastic.

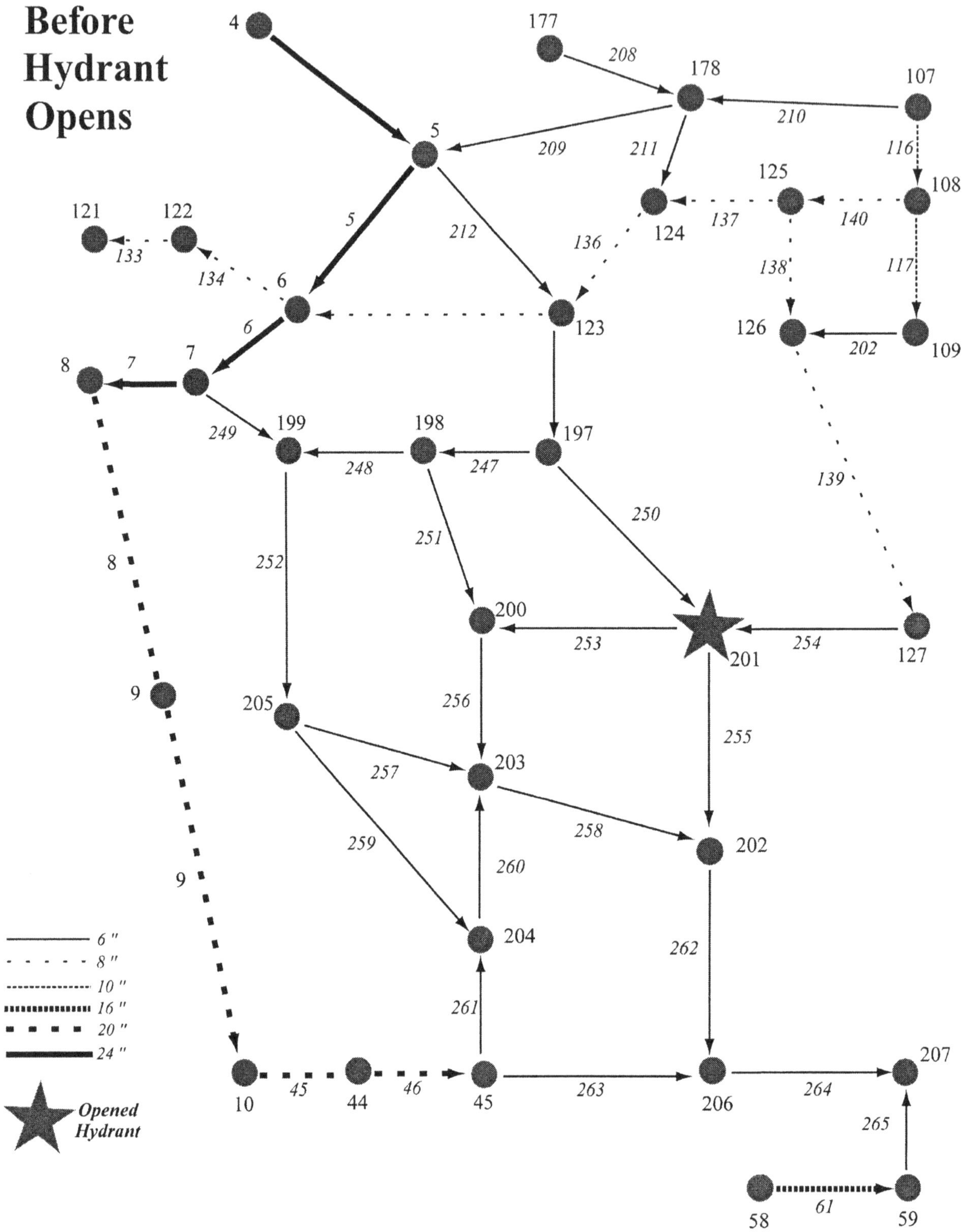

Figure 4.11 Schematic of nodes and pipes in the vicinity of Node #201 before hydrant opens (Model City B)

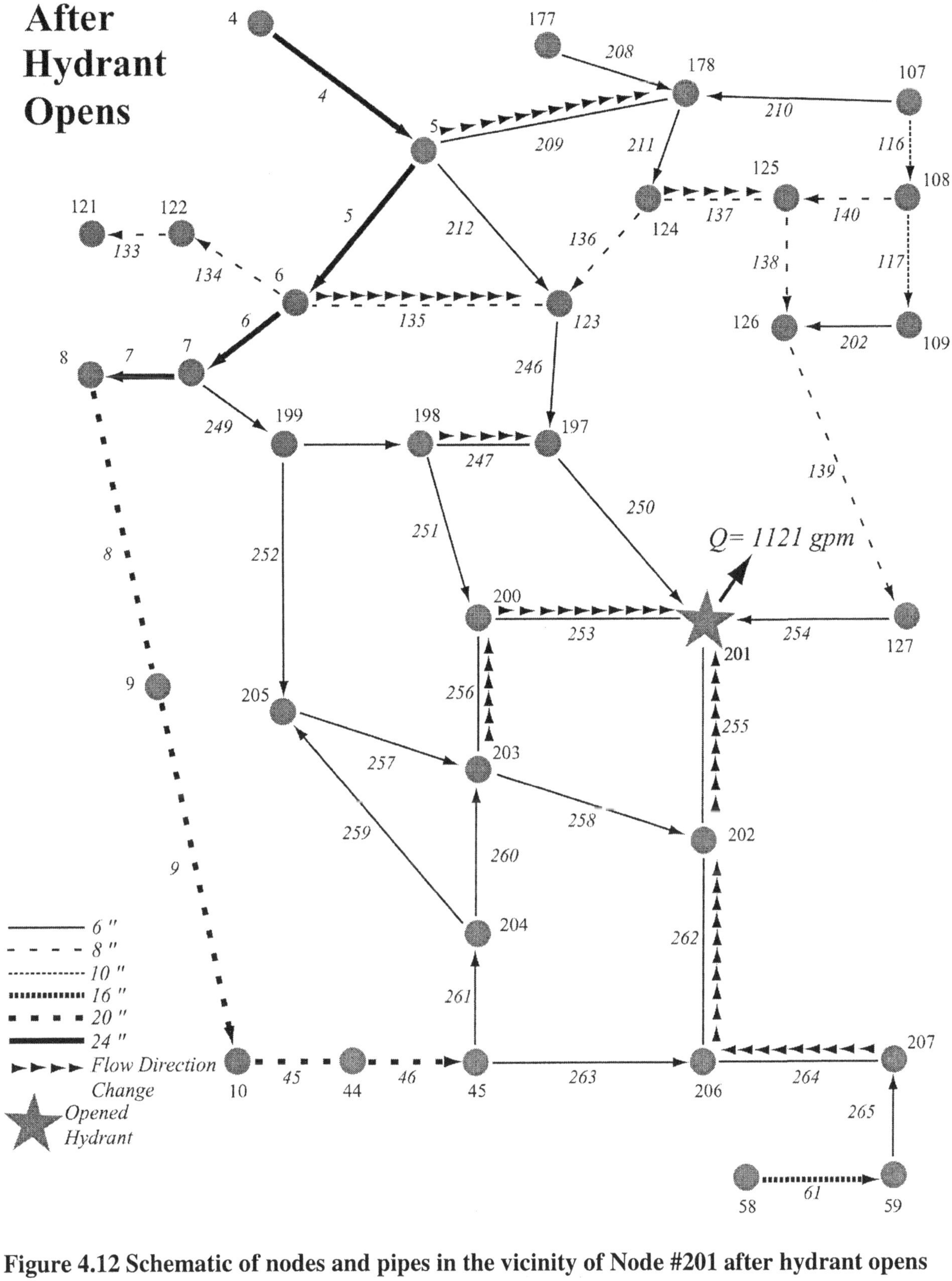

Figure 4.12 Schematic of nodes and pipes in the vicinity of Node #201 after hydrant opens (Model City B)

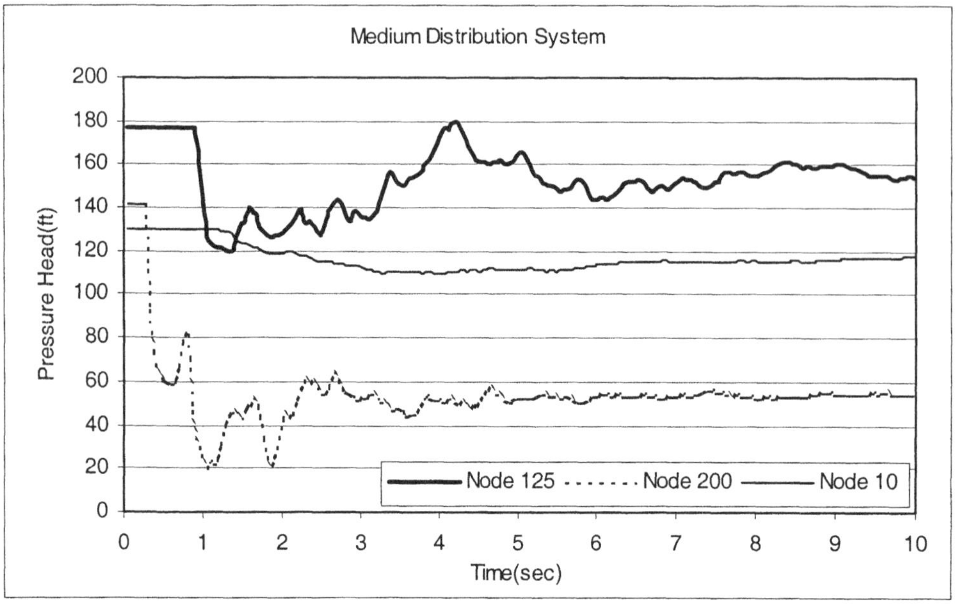

Figure 4.13 Transient pressure head fluctuations at Node #125, #200, #10 (Model City B, Simulation No.1)

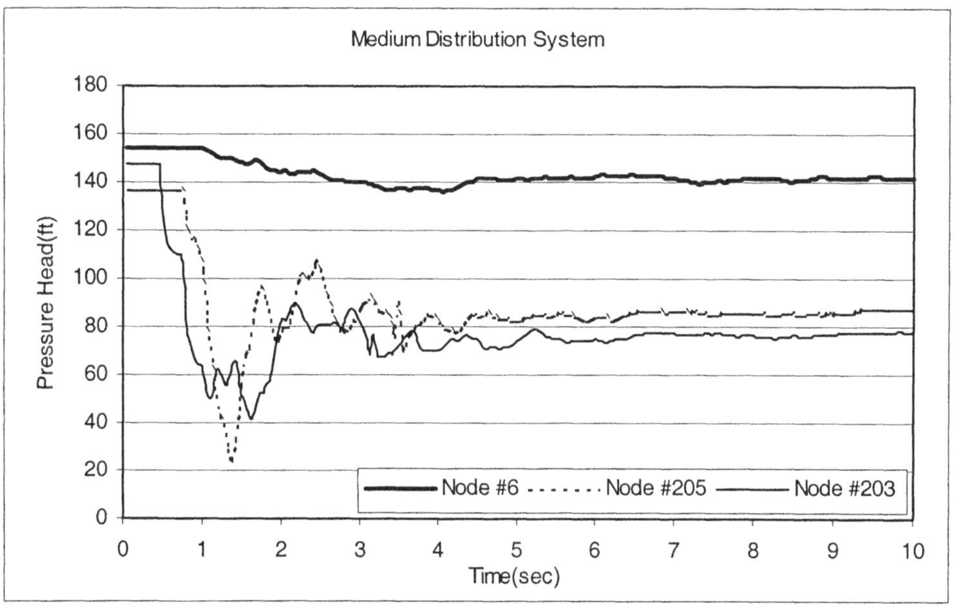

Figure 4.14 Transient pressure head fluctuation at Node #6, #205, #203 (Model City B, Simulation No.1)

The dramatic changes in flows at some of the pipes caused by the opening of a hydrant at Node #201 are shown in Figures 4.15 and 4.16. From Figure 4.15, it is seen that the flow in Pipe #253 (connecting Node #200 and #201) reverses very dramatically almost immediately after the hydrant begins to open. In examining Figure 4.16, it is seen that the flow at Pipe #46 (located near the lower left corner in Figure 4.11) increases dramatically in order to meet the sudden change in demand at Node #201. Flow direction at Pipe #209 (located at the top in Figure 4.11) and Pipe #264 (located near the lower right corner in Figure 4.11) are also reversed to meet the sudden demand at Node #201.

Simulation No. 2 (Model City B)

In order to examine the effect of changes in ground elevation on the transient pressure in the pipe and nodes as well as flows in the pipe, Model City B Simulation No. 2 was performed. Simulation No. 2 is performed with an assumed minor modification of ground elevation at three locations (as compared to Simulation No. 1).

Table 4.5
Ground elevation at selected Nodes for Simulations No. 1 and No. 2

Node	Ground Elevation (ft) Simulation No. 1	Ground Elevation (ft) Simulation No. 2
#198	96.0	110.9
#201	96.4	90.4
#203	96.9	110.0

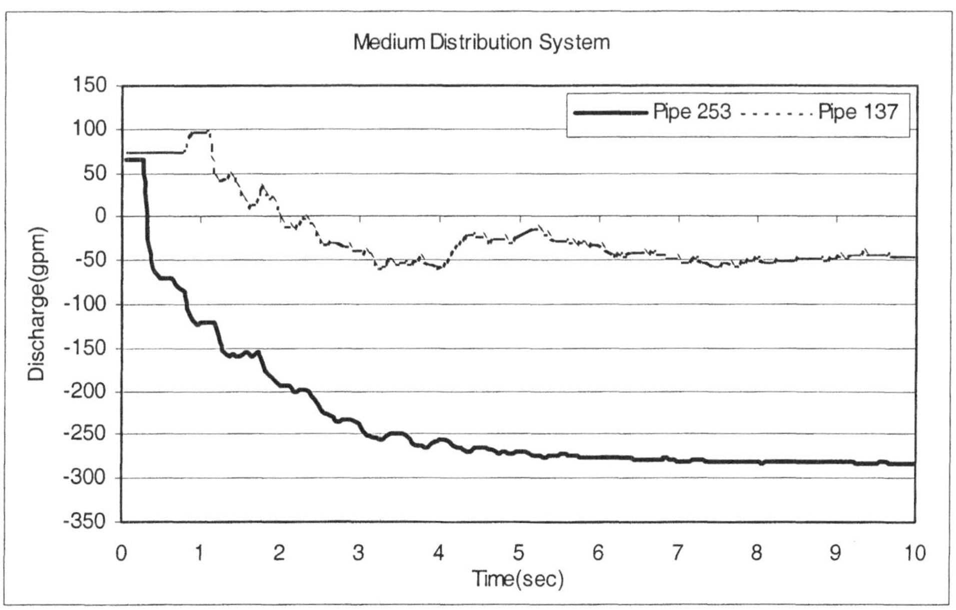

Figure 4.15 Transient flow rates at Pipe #253, #137 (Model City B, Simulation No.1)

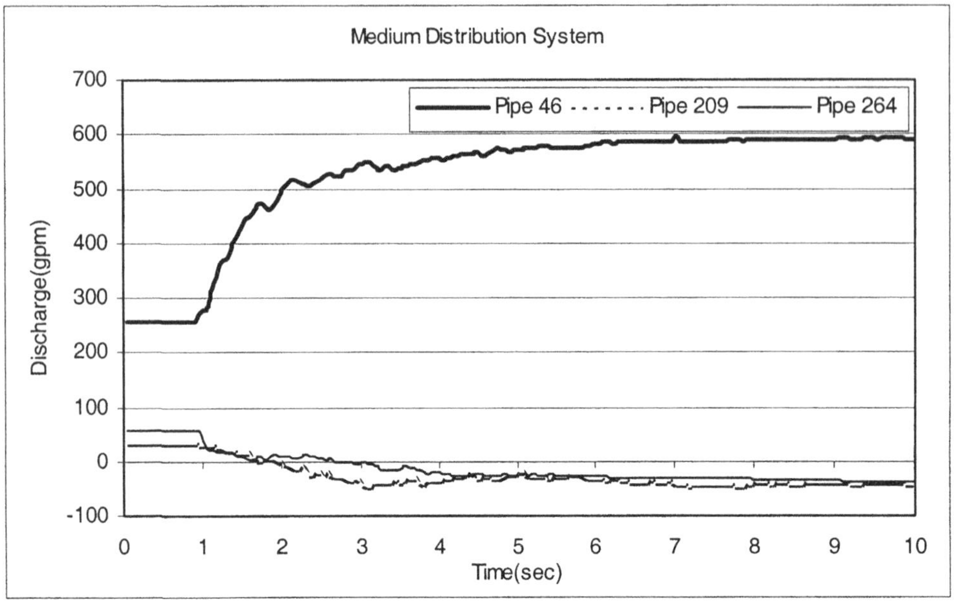

Figure 4.16 Transient flow rates at Pipe #46, #209, #164 (Model City B, Simulation No.1)

Thus, the only difference between Simulation No. 1 and Simulation No. 2 is that the ground elevation at the neighboring area of hydrant at Node #201 is modified. All other conditions remain the same.

Figure 4.17 shows the transient pressure fluctuation at Node #125, #200, and #10 for Simulation No. 2. The total pressure head drop is more than 120ft, furthemore, the low pressure at the trough point is at the undesirable –20ft (pulling a vacuum). By comparing Figure 4.17 with Figure 4.13 one finds that the general pattern of the pressure fluctuation at the three respective locations is quite similar. The absolute value at Node #200 is different and that the negative pressure exists for Simulation No. 2 (Figure 4.17) but not for Simulation No.1. This demonstrates that had the ground level as prescribed by Simulation No. 2 existed, the negative pressure below atmospheric pressure would be the result at Node#200 when drastic demand change at Node #201 occurs.

Figure 4.18 presents the transient pressure fluctuation at Node #6, #205, and #203 (Model City B, Simulation No.2). The dramatic changes in pressure head are evident for Node #205 and #203 although, at the low pressure point the pressure heads at both locations are still positive. By comparing the results in Figure 4.18 with that in Figure 4.14, one sees that the general pressure pattern is quite similar in each of the respective three locations but the absolute value is different.

Figure 4.19 shows the transient flow rate at Pipe #253 and #137. Flow at Pipe #253 reverses quickly to meet the change in demand as a result of the hydrant opening at Node #201. Again the general pattern is similar to that shown in Figure 4.15. A similar curve for Pipe #46, #209, and #264 is shown in Figure 4.20 showing similar trend as that presented in Figure 4.16. The result of Simulation No. 2 (allowing more variation of ground elevations) shows that the transient pressure can reach negative pressure range easily when the hydrant opens suddenly. Such undesirable low-pressure condition can cause back flow and backsiphonage, thus, worsen the cross-connection problems.

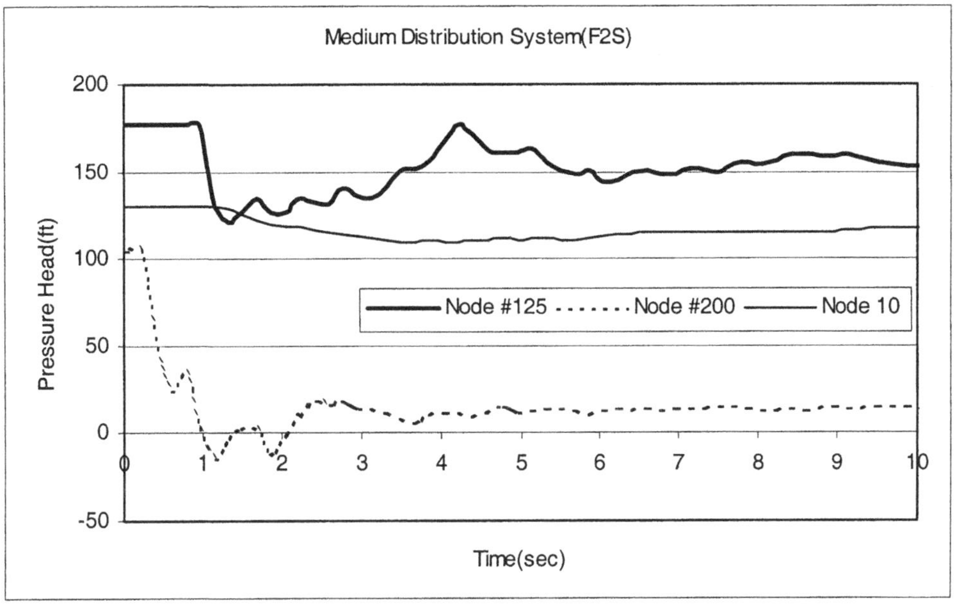

Figure 4.17 Transient pressure head fluctuations at Node #125, #200, #10 (Model City B, Simulation No. 2, hydrant opens in 2 seconds)

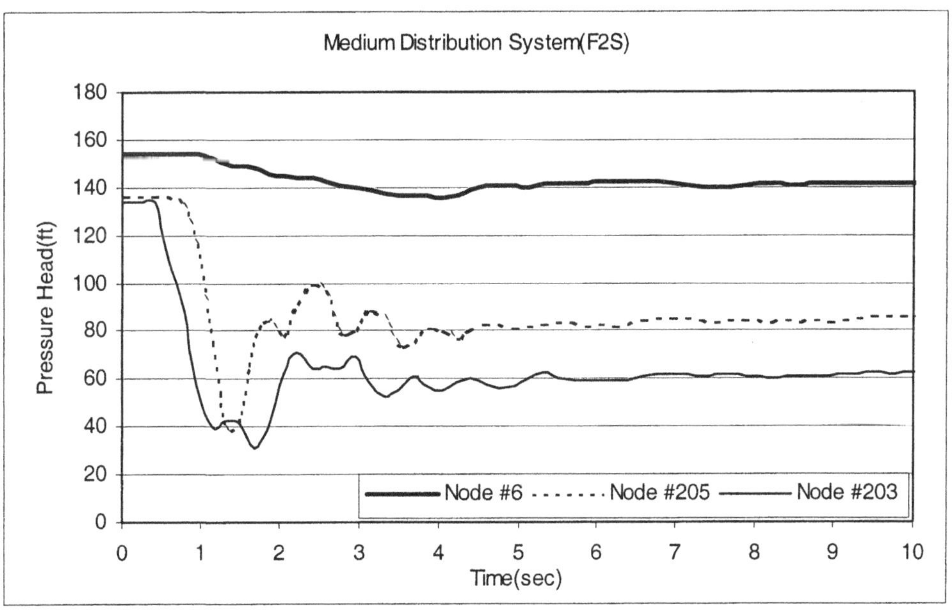

Figure 4.18 Transient pressure head fluctuations at Node #6, #205, #203 (Model City B, simulation No. 2, hydrant opens in 2 seconds)

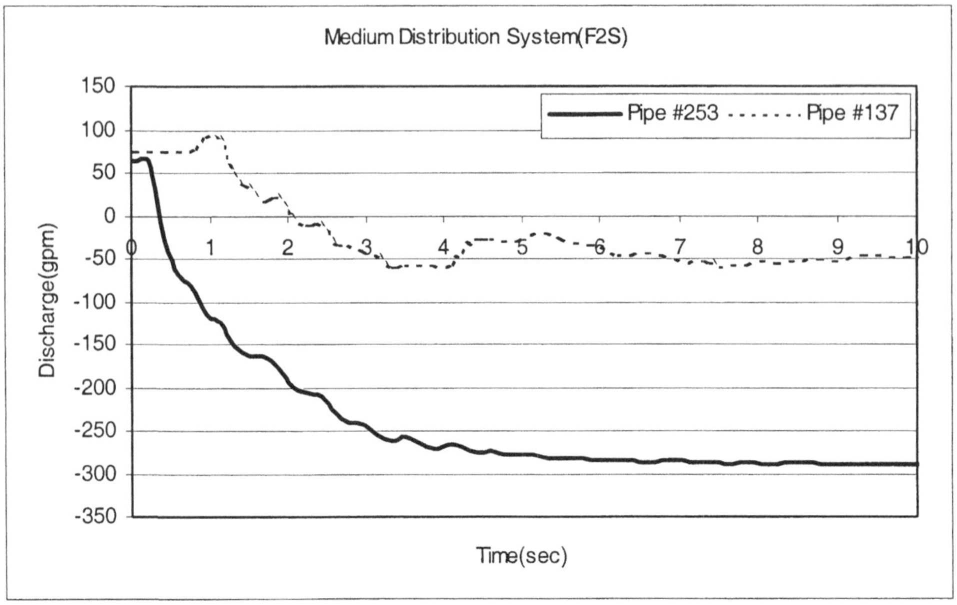

Figure 4.19 Transient flow rates at Pipe #253 and #137 (Model City B, Simulation No. 2, hydrant opens in 2 seconds)

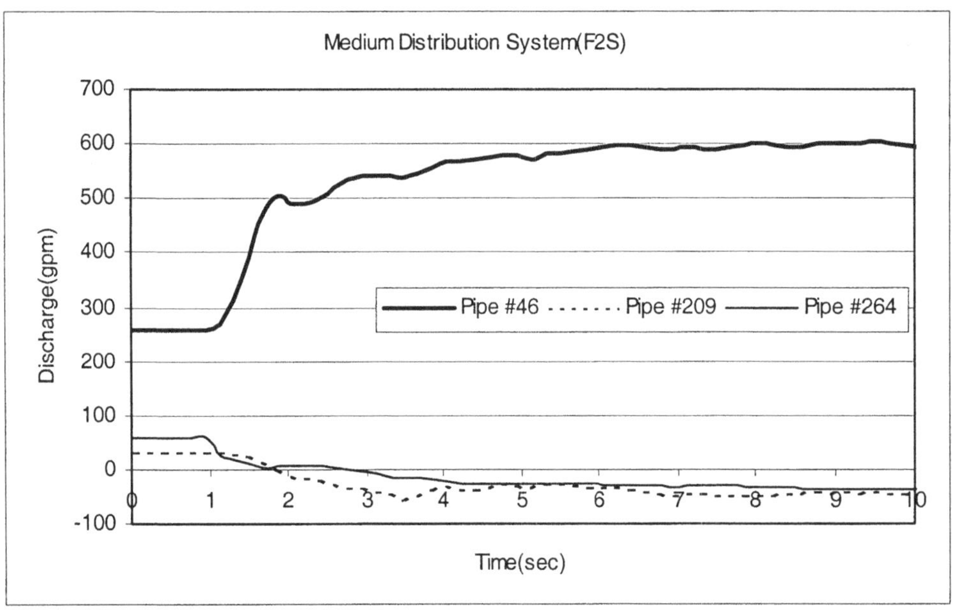

Figure 4.20 Transient flow rates at Pipe #46, #209, #264 (Model City B, Simulation No. 2, hydrant opens in 2 seconds)

Simulation No. 3 (Model City B)

Another simulation condition for Model City B is conducted for Simulation No. 3. In this simulation, the hydrant at Node #201 is assumed to open in 5 seconds. Thus, the only difference between Simulation No.2 and Simulation No. 3 is that in Simulation No. 2 the hydrant opens in 2 seconds, but in Simulation No. 3 the hydrant opens in 5 seconds. This simulation is to ascertain the effect of slower opening of the hydrant. Figure 4.21 presents the pressure fluctuation at Node #125, #200, and #10 when the hydrant at Node #201 opens in 5 seconds. From Figure 4.21, it is seen that the pressure drop at Node #200 is somewhat less and that the low pressure point does not reach the negative value when compared with Figure 4.17. Figure 4.22 shows the pressure at Node #6, #205, and #203. Again, the low pressure point does not reach the level shown in Figure 4.18 confirming the slower hydrant-opening rate does lessen the dramatic pressure drop even though the general pattern is quite similar. Figures 4.23 and 4.24 show the discharge rate at five pipes when the hydrant opens in 5 seconds. Results shown in these figures confirm that the slower hydrant-opening rate will lessen the impact of hydraulic transient (pressure and discharge) caused by the sudden changes in demand such as opening the hydrant.

Comparison of pressure head fluctuation at three nodes: Node #200, #205, and #202 for different hydrant opening rates (2 sec. vs. 5 sec.) are presented in Figure 4.25, 4.26, and 4.27 respectively. By reviewing these figures, it is clear that when the hydrant opens in 2 seconds the resulting transient flow at the vicinity of the open hydrant will cause a larger reduction in pressure. In some situations, such as that at Node #200, negative pressure is evident.

Model City A (Large Network System)

To represent a large sized system, we used an existing distribution network in the southwest region of the united states. The system is supplied by a pump station located at a water treatment plant that delivers 236cfs to meet the system demands and six storage reservoirs. Most of the nodes identified in the system are demands that take flow from the system. Figures 4.28 and 4.29 show the schematics of a part of the system in the immediate vicinity of Node #181 before and after the hydrant at Node #181 opens. This large network system is one order of magnitude larger than the medium-sized system (Model City B). Pipes in this system are bigger and distance between nodes (pipe length) is longer. The system consists of 231 nodes and 321 pipes. The condition at steady state before the transient analysis begins is provided in Appendix 6. In this appendix, data associated with the node elevation, discharge, pipe length, pipe diameter, and the connectivity to each node are also provided.

Simulation No. 1 (Model City A, hydrant open in 2 seconds)

For the transient analysis, it was assumed that a hydrant at Node #181 opened in two seconds (Simulation No. 1), creating a sudden demand of 4985gpm at this location. Because of the large-sized pipelines in this model, the flow was limited to this amount instead of assuming a full capacity opening which would result in flows in excess of 11,000gpm. The sudden demand caused by the opening of hydrant at Node #181 caused many flow reversals in the system. The additional dash arrow lines in Figure 4.29 for some of the pipes indicate the pipes experiencing the flow reversal. Some of these flow reversals occur at location quite a distance from Node

#181. From Figure 4.29, it is seen that one of the pipeline flow reversals occurs on the line to a reservoir and another occurs in a 48-inch diameter pipe. Four different simulation runs are conducted for this large network system (Model City A).

The results indicated in Figure 4.29 are from Simulation No. 1. The hydrant opening occurs at Node #181 and is completed in two seconds, reaching a flow rate of 4985gpm. Figure 4.30 presents the transient pressure fluctuation at Node #177, #111, and #116 showing the transient effect at locations with different distances from the open hydrant. At Node #177, located 600ft from the hydrant, the computed pressure will drop below atmospheric pressure for a brief instant with an overall pressure drop of 125ft. At node #111, some distance away from the hydrant location, the pressure fluctuation is less dramatic, as would be expected, but is still significant. At Node #116, further away from the hydrant location the magnitude of fluctuating pressure is smaller, but cannot be ignored. These pressure fluctuations clearly indicate that a sudden demand, such as the opening of a fire hydrant, can have far reaching effects on the distribution system, just as was the case with the medium sized system discussed in the previous section.

Transient pressure fluctuation at three other locations, Node #190, #182, and #107 is presented in Figure 4.31. The pressure at Node #182 is expected to drop over 90ft although it does not drop below atmospheric pressure. The locations in these figures are chosen with different distances from the open hydrant. The transient flow rate at Pipe #237, #195, and #320 is presented in Figure 4.32. It is seen that in all of these three pipes there is flow reversal due to the open hydrant at Node #181. The flow rate fluctuation at three other pipes, Pipe #201, #233, and #245 is shown in Figure 4.33 showing the varying degree of impact on the flow rate depending on the distance away from the open hydrant location.

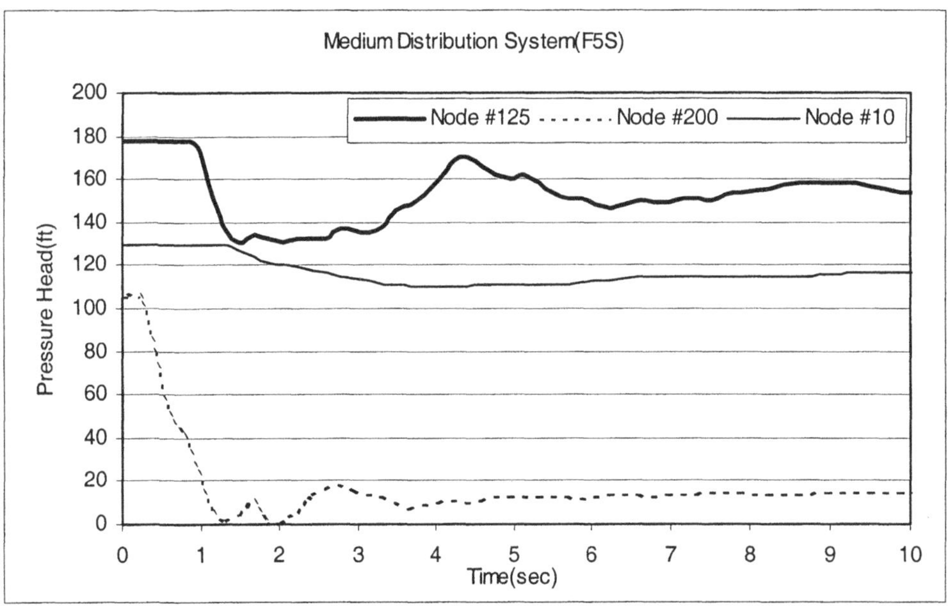

Figure 4.21 Transient pressure head fluctuations at Node #125, #200, #10 (Model City B, Simulation No. 3, hydrant opens in 5 seconds)

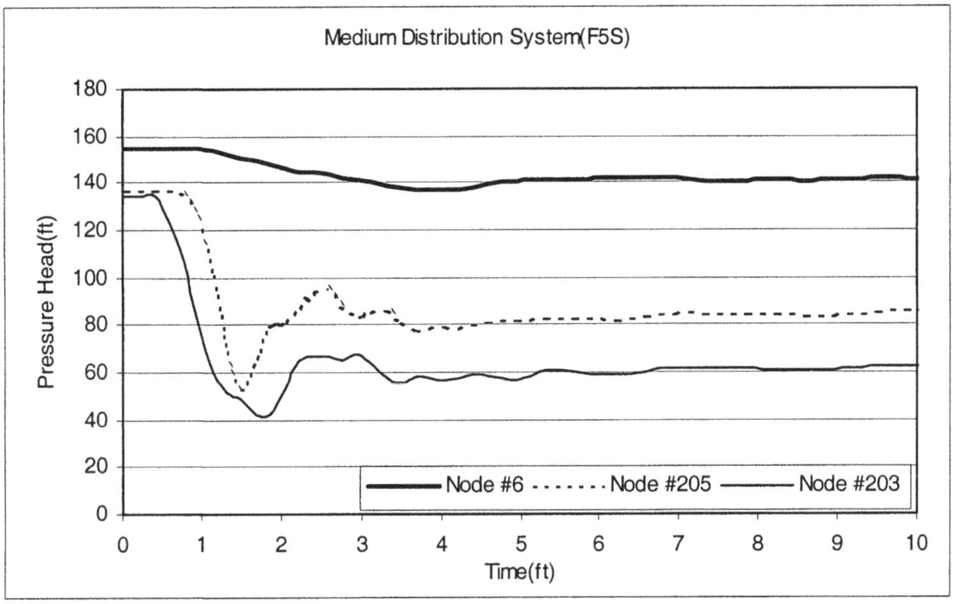

Figure 4.22 Transient pressure head fluctuations at Node #6, #205, #203 (Model City B, Simulation No. 3, hydrant opens in 5 seconds)

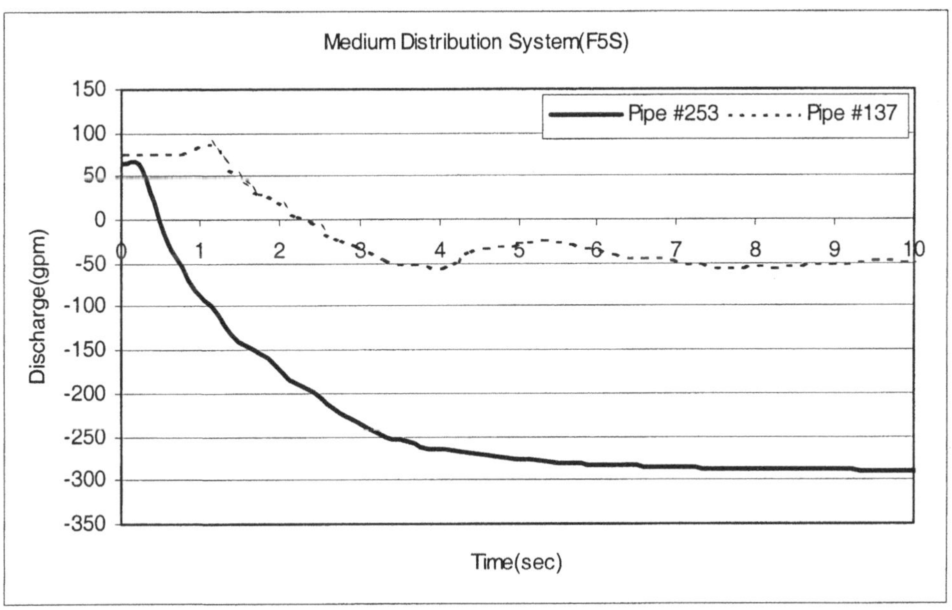

Figure 4.23 Transient flow rates at Pipe #253 and #137 (Model City B, Simulation No. 3, hydrant opens in 5 seconds)

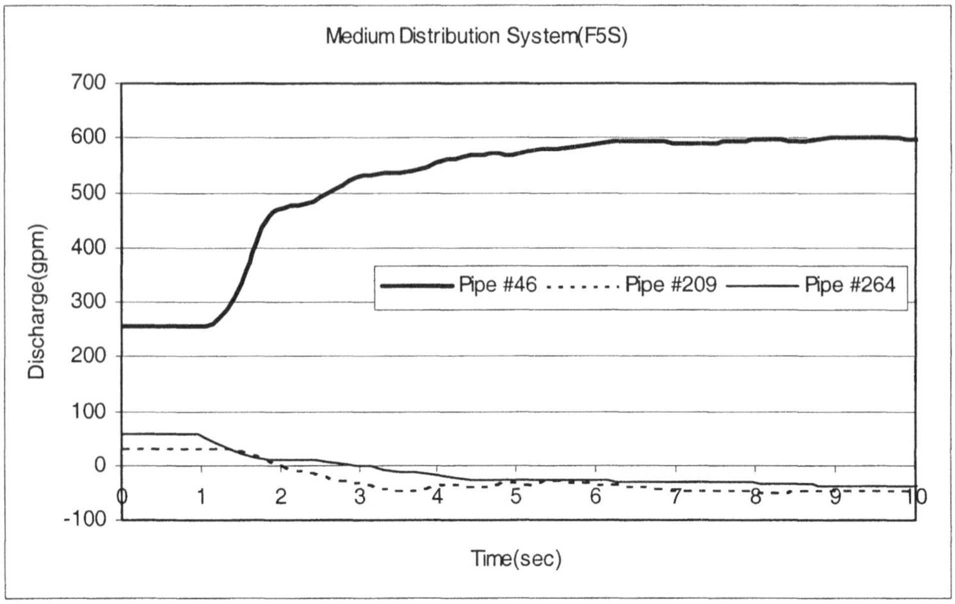

Figure 4.24 Transient flow rates at Pipe #46, #209, #264 (Model City B, Simulation No. 3, hydrant opens in 5 seconds)

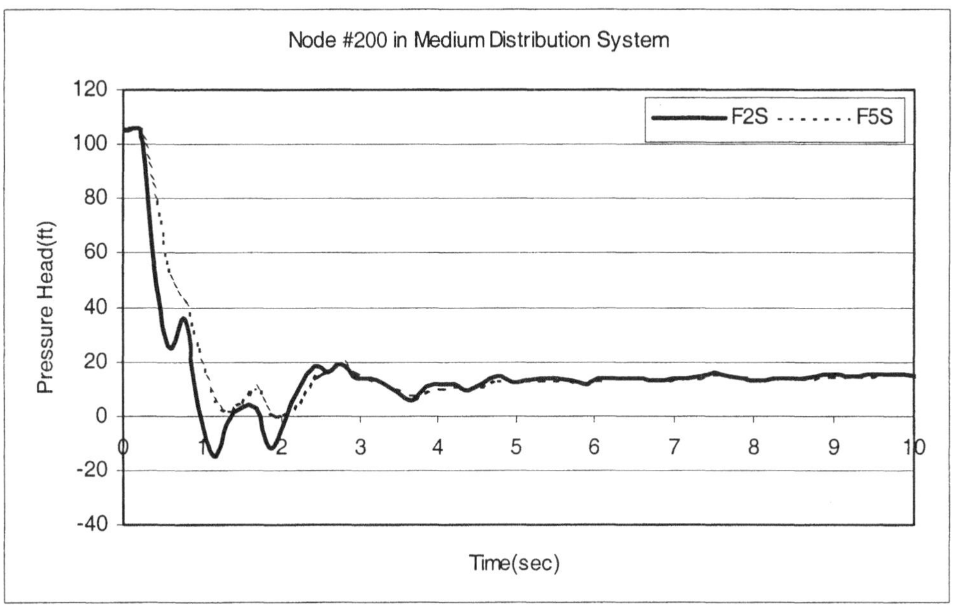

Figure 4.25 Comparison of pressure head at Node #200 for different hydrant opening rates (2sec vs. 5 sec)

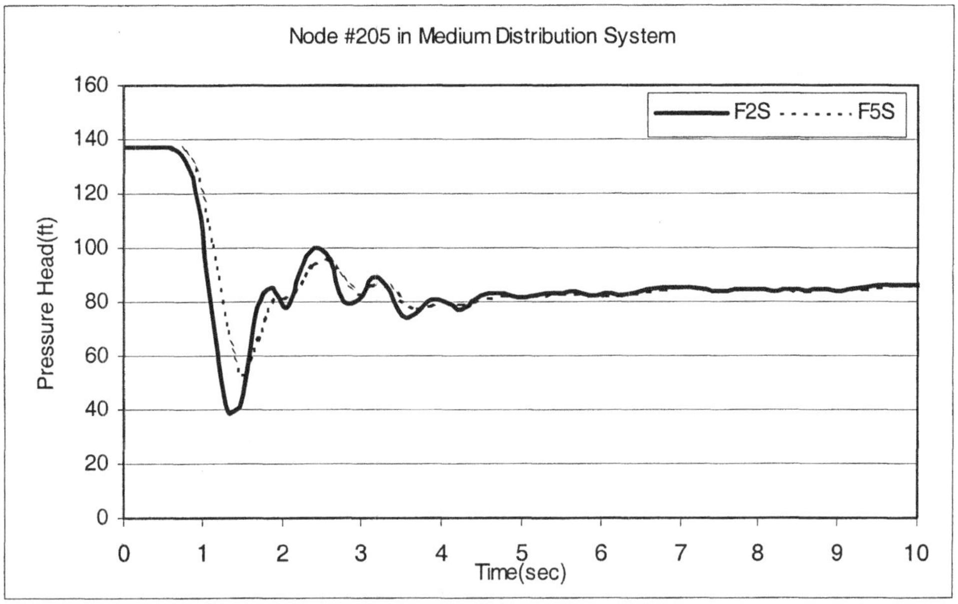

Figure 4.26 Comparison of pressure head of Node #205 for different hydrant opening rates (2 sec vs. 5 sec)

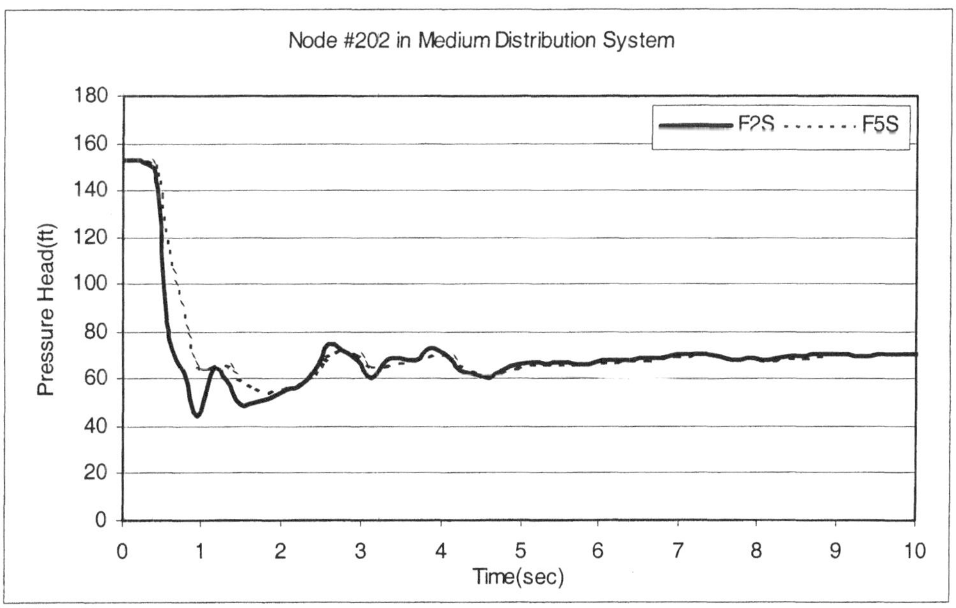

Figure 4.27 Comparison of pressure head at Node #202 for different hydrant opening rates (2 sec vs. 5 sec)

Before Hydrant Opens

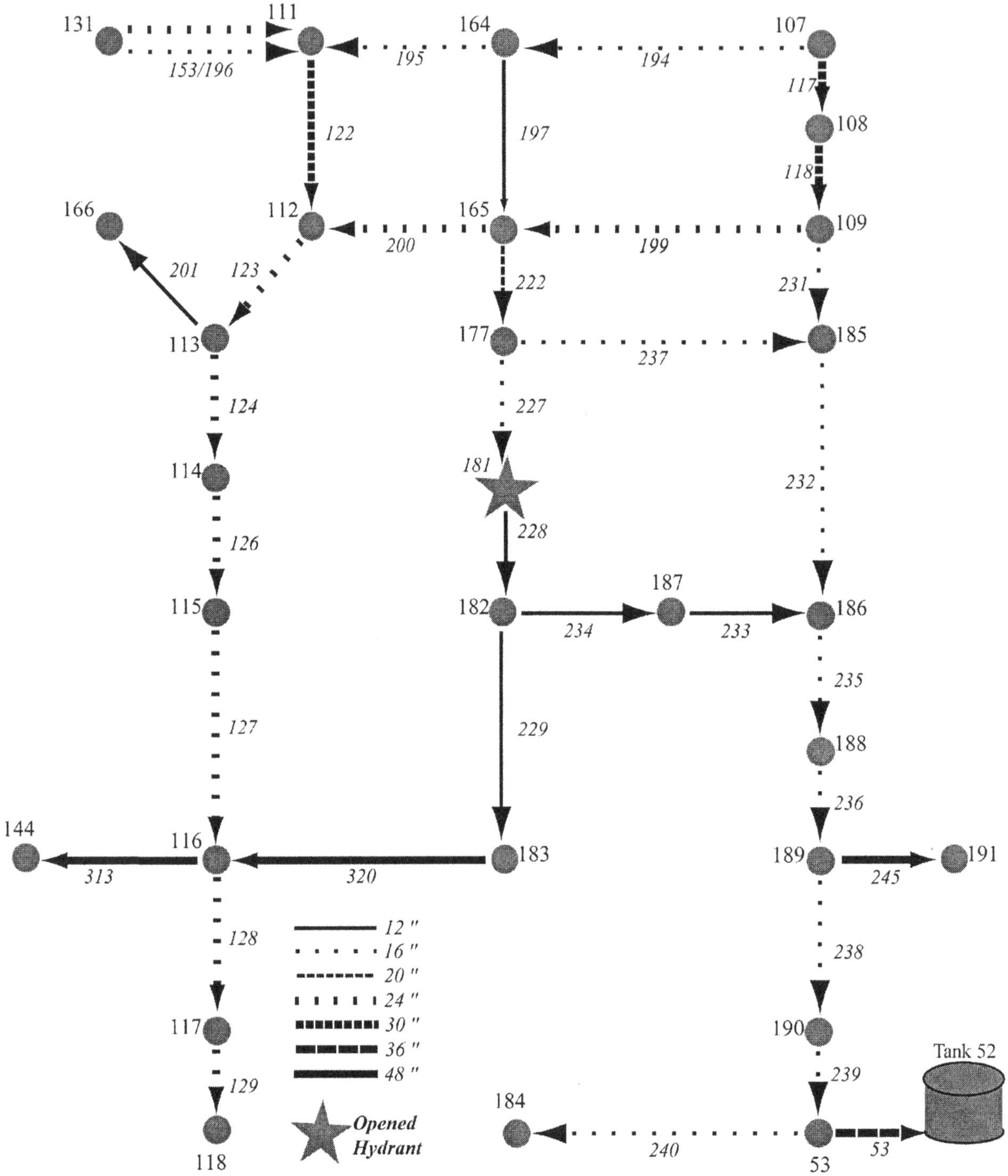

Figure 4.28 Schematic of nodes and pipes in the vicinity of Node #181 before hydrant opens (Model City A)

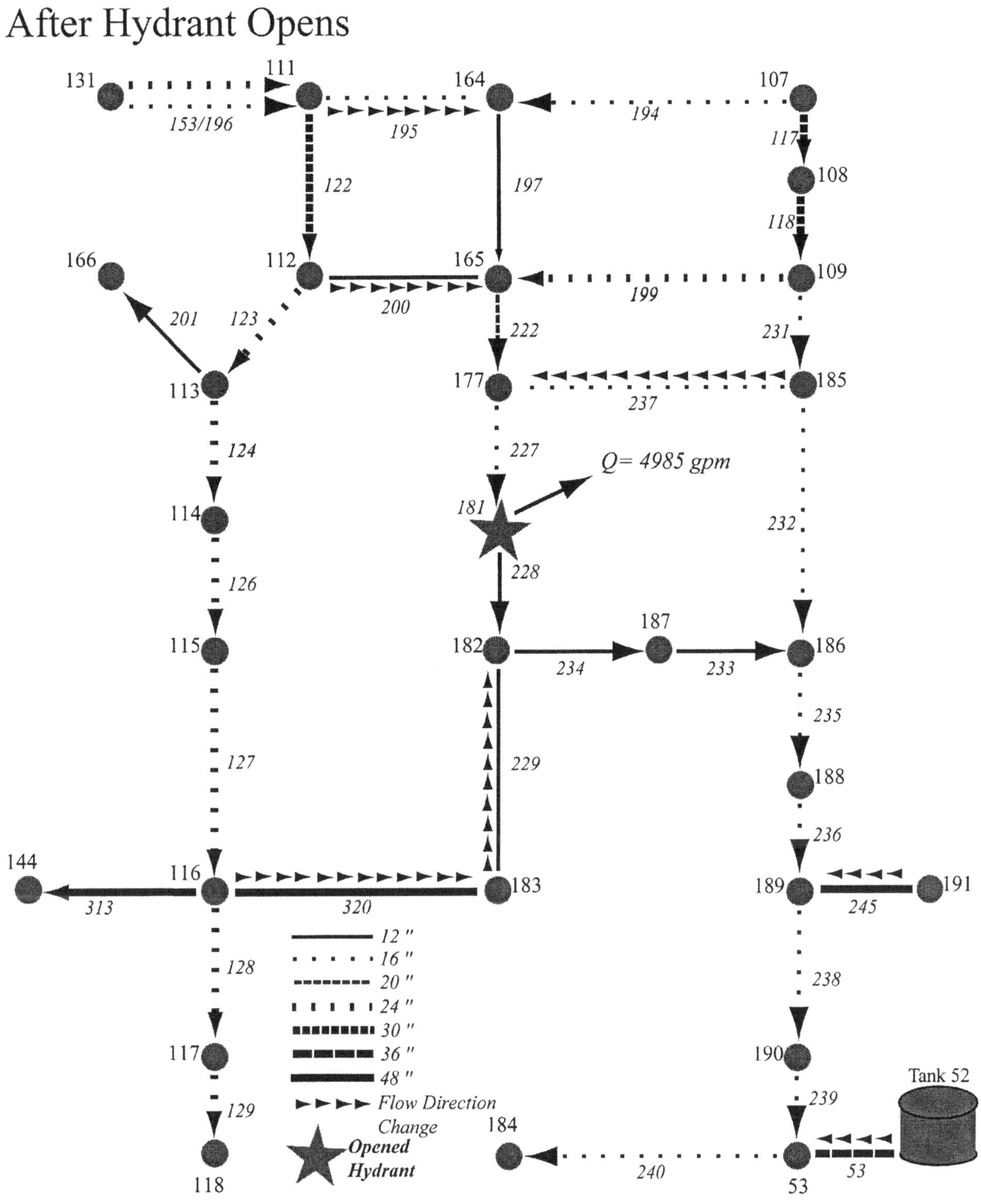

Figure 4.29 Schematic of nodes and pipes in the vicinity of Node #181 after hydrant opens (Model City A)

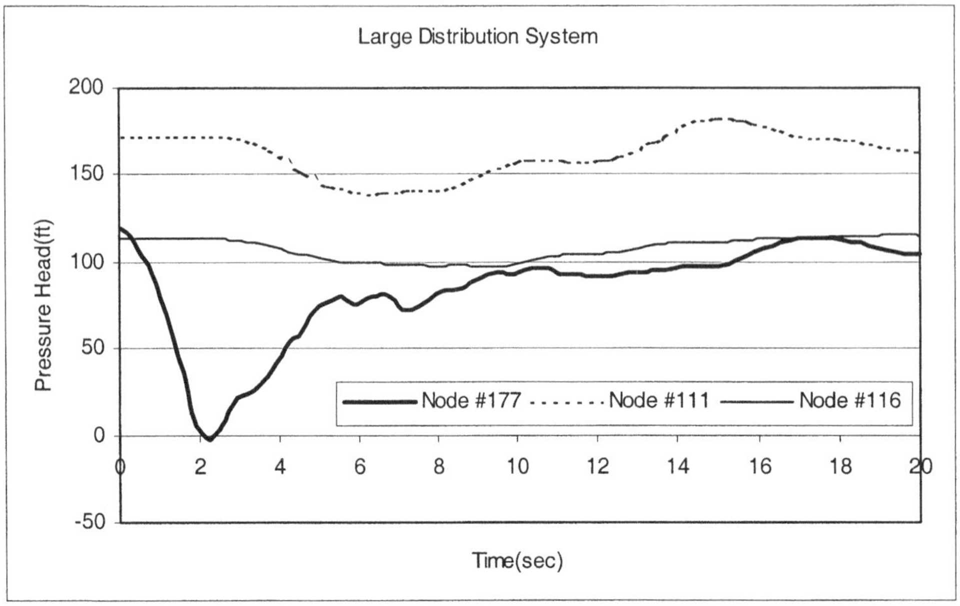

Figure 4.30 Transient pressure head fluctuations at Node #177, #111, #116 (Model City A, Simulation No. 1, hydrant opens in 2 seconds)

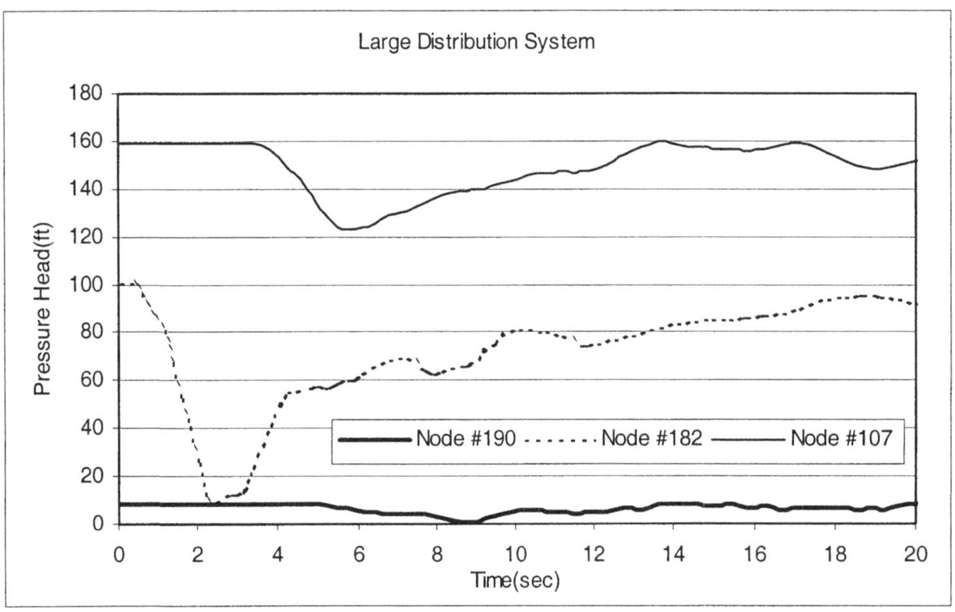

Figure 4.31 Transient pressure head fluctuations at Node #190, #182, #107 (Model City A, Simulation No. 1, hydrant opens in 2 seconds)

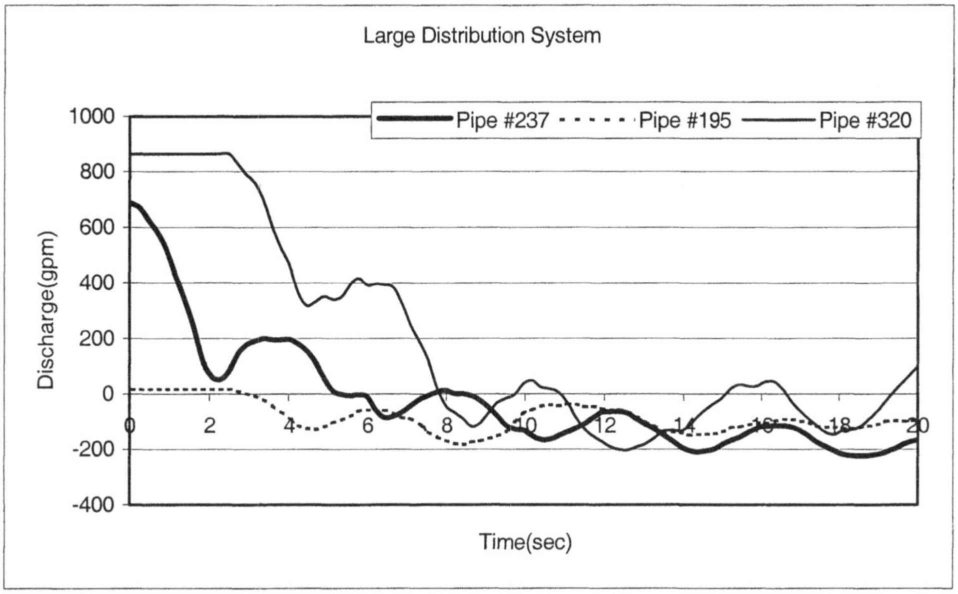

Figure 4.32 Transient flow rates at Pipe #237, #195, #320 (Model City A, Simulation No. 1, hydrant opens in 2 seconds)

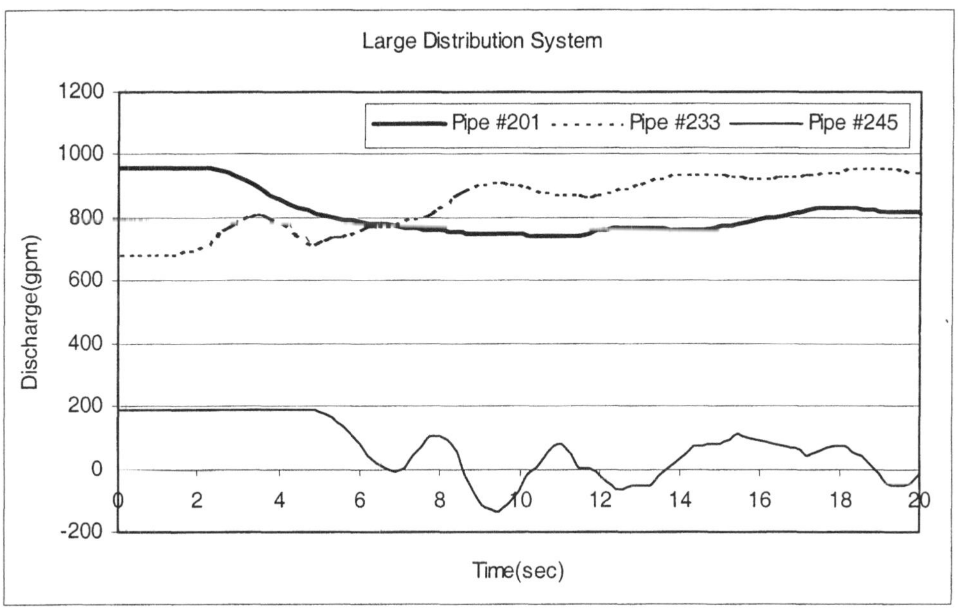

Figure 4.33 Transient flow rates at Pipe #201, #233, #245 (Model City A, Simulation No. 1, hydrant opens in 2 seconds)

Simulation No. 2 (Model City A, hydrant open in 2 seconds)

For simulation No. 2 in Model City A (large sized system) the elevation at Node #177 is assumed to be at 750ft (raised by 20ft). All other parameters are the same as in Simulation No. 1: The hydrant opening time is also assumed to be two seconds. The pressure fluctuation at Node #177, #111, and #116 is presented in Figure 4.34. The low pressure point at Node #177 clearly drops below atmospheric pressure, until it gradually recovers after three second. Thus, the results clearly show that the pressure drops at locations close to the open hydrant are very dramatic. Whether pressure will drop below atmospheric pressure (pulling vacuum) will certainly depend on the elevation of the node. Sudden drops of pressure, especially when dropping below atmospheric pressure will be a serious problem if direct or in-direct cross-connections exist. The examples shown so far do indicate the real potential for adverse impact. Pressure fluctuation at three other locations Node #190, #182, and #107 is shown in Figure 4.35. The flow rate fluctuations at six pipelines are presented in Figures 4.36 and 4.37. The impact on Pipe #237 is quite dramatic as can be seen in Figure 4.36. Before the transient analysis begins, the discharge of nearly +700gpm is flowing in Pipe #237. As soon as the hydrant open for 2 seconds, the flow in Pipe #237 quickly reverses and reaches a level of –200gpm before the oscillations set in for many cycles. Flows in Pipe #320 experience a similar pattern as well.

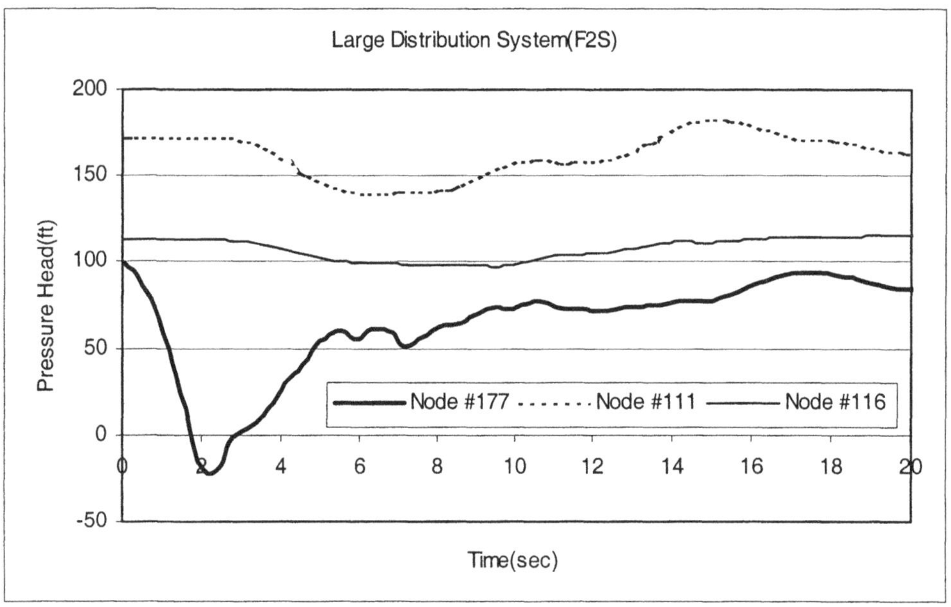

Figure 4.34 Transient pressure head fluctuations at Node #177, #111, #116 (Model City A, Simulation No. 2, hydrant opens in 2 seconds)

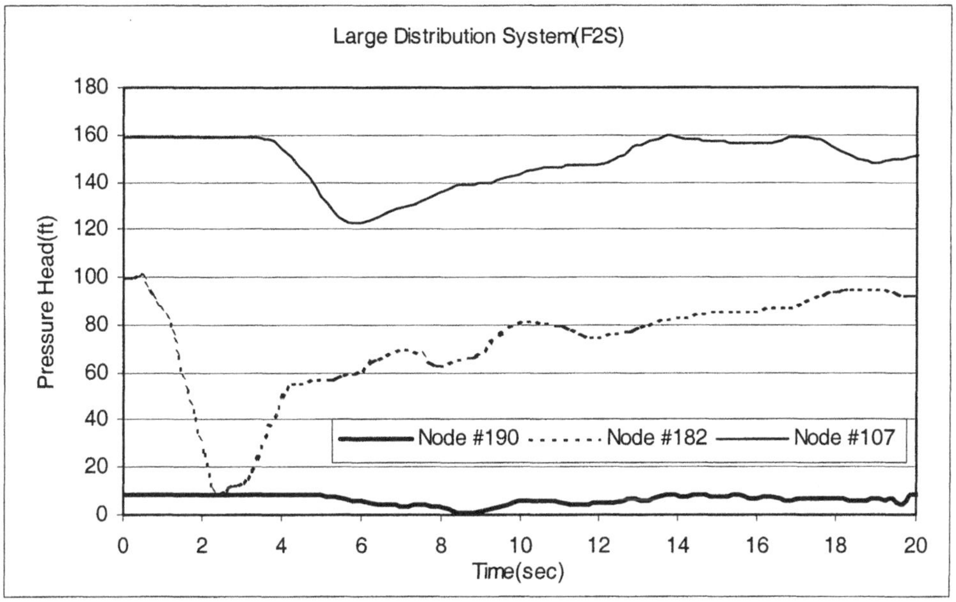

Figure 4.35 Transient pressure head fluctuations at Node #190, #182, #107 (Model City A, Simulation No. 2, hydrant opens in 2 seconds)

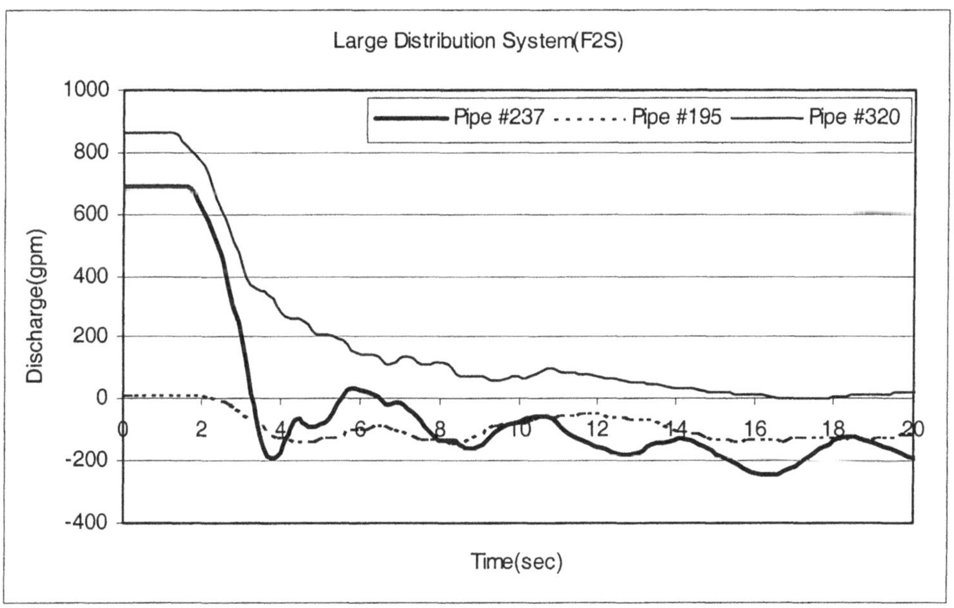

Figure 4.36 Transient flow rates at Pipe #237, #195, #320 (Model City A, Simulation No. 2, hydrant opens in 2 seconds)

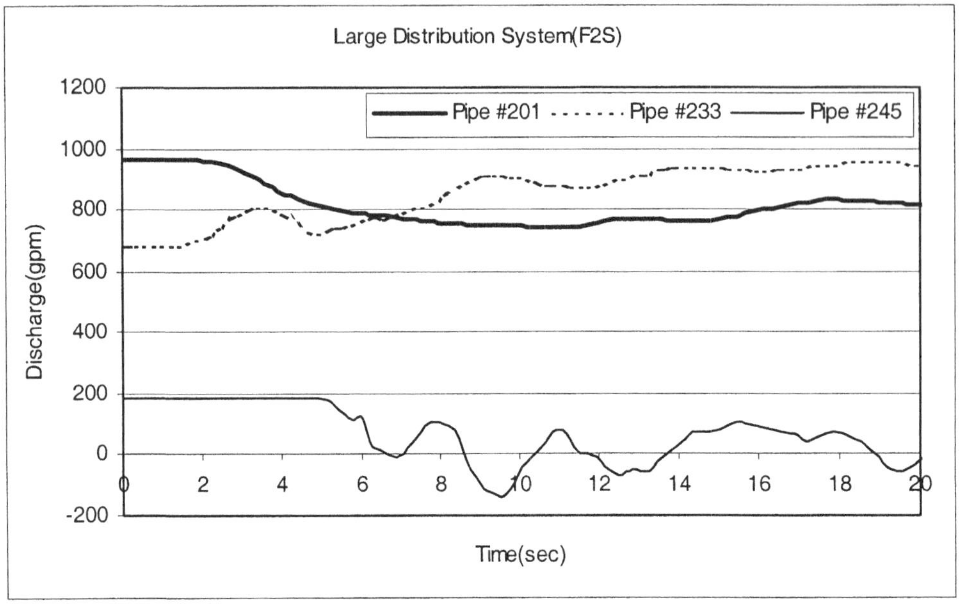

Figure 4.37 Transient flow rates at Pipe #201, #233, #245 (Model City A, Simulation No. 2, hydrant opens in 2 seconds)

Simulation No. 3 (Model City A, hydrant opens in 5 seconds)

The hydrant opening time is changed from two seconds to five seconds in this simulation. Thus, the only difference between Simulation No. 3 and Simulation No. 2 is the hydrant opening time. The four corresponding curves on pressures and flow rates are presented in Figures 4.38 to 4.41. Looking at Figure 4.38, the drop in pressure at Node #177 is not as rapid as that shown in Figure 4.34, the low pressure point is just barely reaching atmospheric pressure. Therefore, by increasing the hydrant opening time, it goes a long way to lessen the impact of the resulting hydraulic transients caused by the change in demands at certain junctions. The same observation can also be made by comparing Figure 4.39 with 4.35; Figure 4.40 with 4.36; and Figure 4.41 with Figure 4.37. The longer hydrant opening time will greatly lessen the impact of the sudden change in pressure and flow.

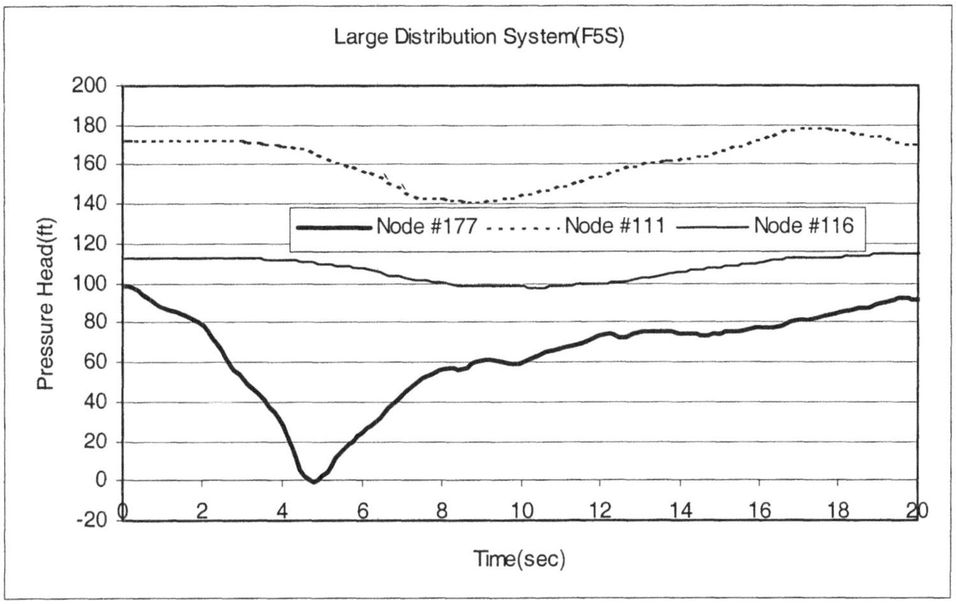

Figure 4.38 Transient pressure head fluctuations at Node #177, #111, #116 (Model City A, Simulation No. 3 hydrant opens in 5 seconds)

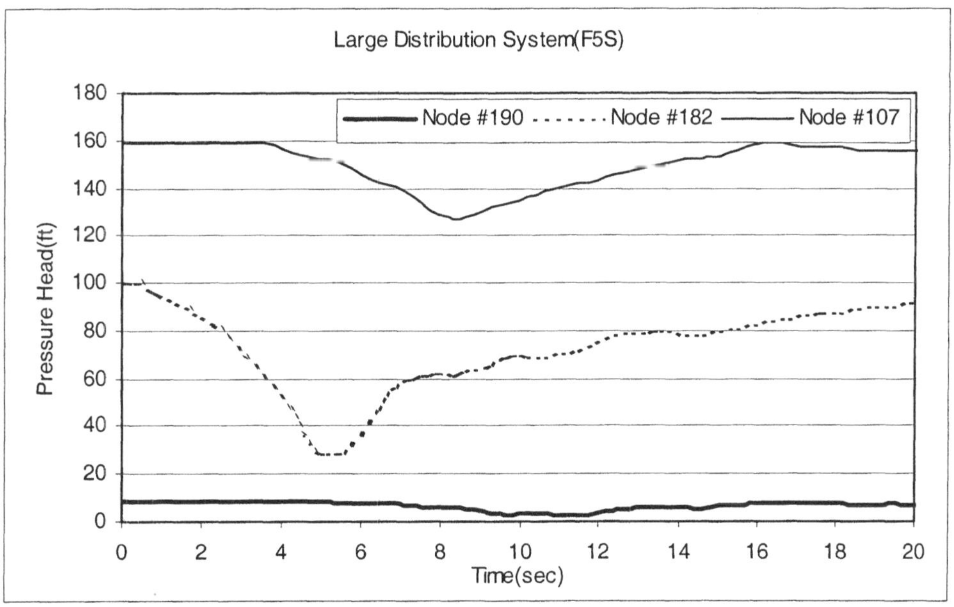

Figure 4.39 Transient pressure head fluctuations at Node #190, #182, #107 (Model City A, Simulation No. 3, hydrant opens in 5 seconds)

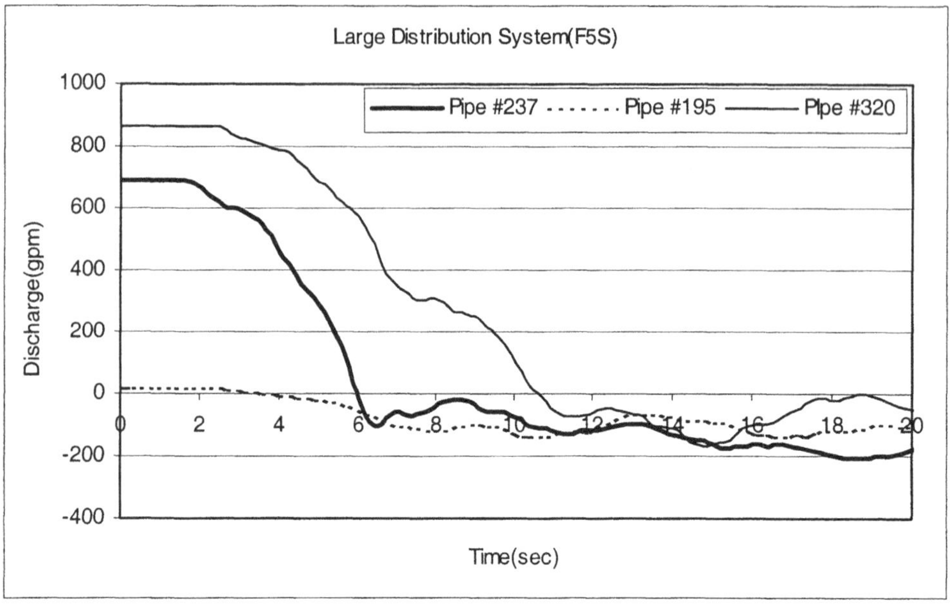

Figure 4.40 Transient flow rates at Pipe #237, #195, #320 (Model City A, Simulation No. 3, hydrant opens in 5 seconds)

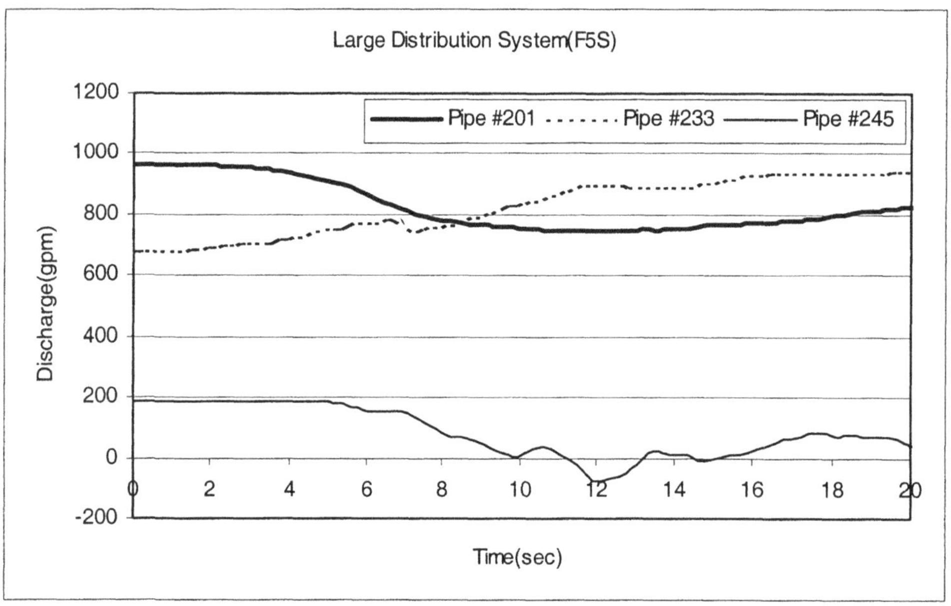

Figure 4.41 Transient flow rates at Pipe #201, #233, #245 (Model City, A, Simulation No. 3, hydrant opens in 5 seconds)

Simulation No. 4 (Model City A, hydrant open in 2 seconds, water level at reservoir raised)

This simulation is performed by raising the water level 60ft, from 817ft to 877ft, at the reservoir Tank 52. Elevation at Node #177 is set at 750ft (the same as in Simulation No.2) and the hydrant opens in two seconds. Thus, the only difference between Simulation No. 4 and Simulation No. 2 is that the water level at the reservoir (Tank 52) is raised by 60ft. The corresponding curves for pressure and flow rate are presented in Figures 4.42 to 4.44. Looking at Figure 4.42, it is seen that the pressure drop at Node #177 does not reach the low point shown in Figure 4.34, even though it also drops below atmospheric pressure. From Figure 4.43, the total pressure drop at Node #182 is approximately 100ft, but because the initial value is now at 120ft, the low pressure point is at 20ft, not as low as that shown in Figure 4.35 (which is approximately 8ft). Similar impact can also be seen in Figure 4.44 when one compares the flow rate time history with those shown in Figure 4.36. Thus, the higher the reservoir water level, the less possibility there is of developing vacuum conditions at low pressure points.

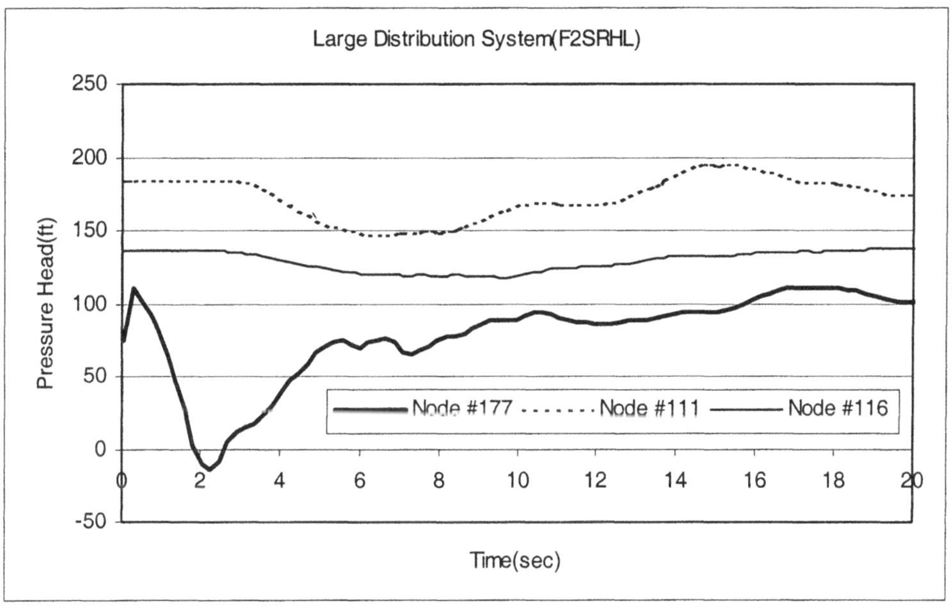

Figure 4.42 Transient pressure head fluctuations at Node #177, #111, #166 (Model City A, Simulation No. 4, hydrant opens in 2 seconds)

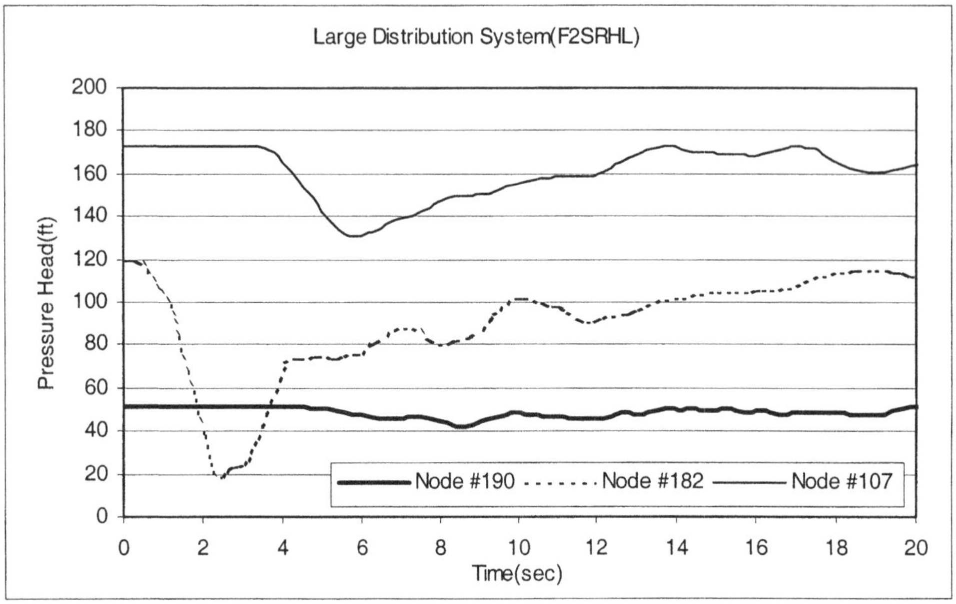

Figure 4.43 Transient pressure head fluctuations at Node #190, #182, #107 (Model City A, Simulation No. 4, hydrant opens in 2 seconds)

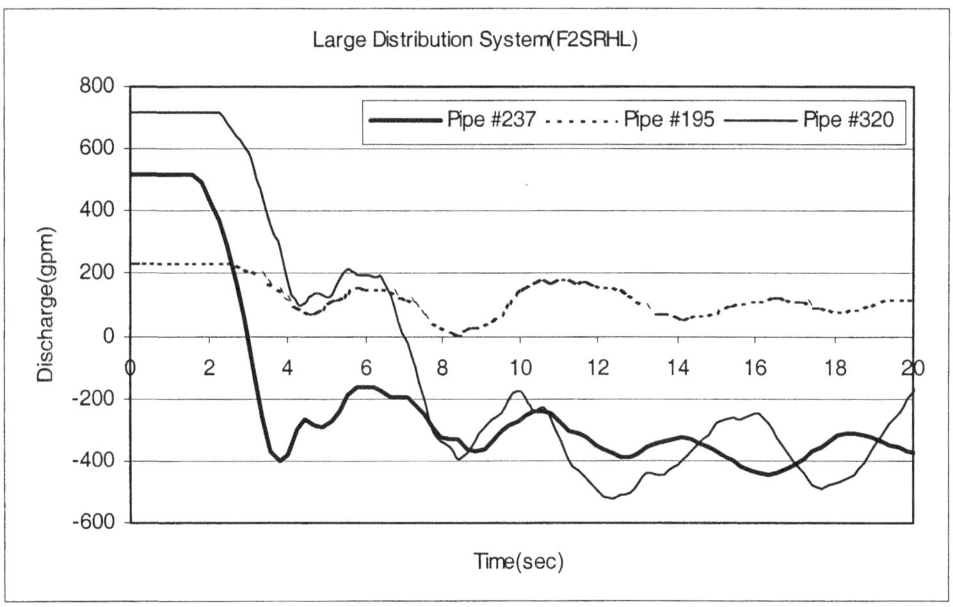

Figure 4.44 Transient flow rates at Pipe #237, #195, #320 (Model City A, Simulation No. 4, hydrant opens in 2 seconds)

COMPARISON OF DIFFERENT SCENARIOS

From the results presented earlier, it is noted that a slower opening rate of the hydrant will lessen the impact of the drastic hydraulic transient phenomena. In order to demonstrate this more clearly, the results on pressure drop in some key locations are presented for direct comparison. Figure 4.45 presents the comparison of pressure head at Node #177 for two different hydrant opening rates: two-seconds vs. five-seconds. From the curves presented, it is clearly seen that for the case of the two-second hydrant opening, the initial pressure to a level below atmospheric pressure occurs rapidly. For the five-second opening, the pressure drop is more moderate and does not go below atmospheric pressure level. Again, this sharp and drastic pressure drop can induce backflow and bring about serious cross-connection related problem into a water distribution system. A similar comparison of pressure drop is presented in Figure 4.46 for Node #182. Again the initial pressure drop at Node #182 is drastic and swift for the two-second hydrant opening while the drop from the five-second opening is more moderate.

The impact on raising the water level at the reservoir is shown in Figure 4.47 for Node #177. The solid black line shows the result of pressure history at Node #177 when the hydrant opening is accomplished in two seconds. The dash line shows the result of raising the reservoir water level by 60ft, with the same two-second opening time for the hydrant. It is seen, from these two curves, that the pressure history follows the same pattern with the elevated water level contributing to uniform increase in the pressure level at Node #177. Figure 4.48 shows the similar comparison at Node #182. Again, the elevated water level at Tank 52 contributes to the general increase in the pressure level and that the same trend is followed for both cases.

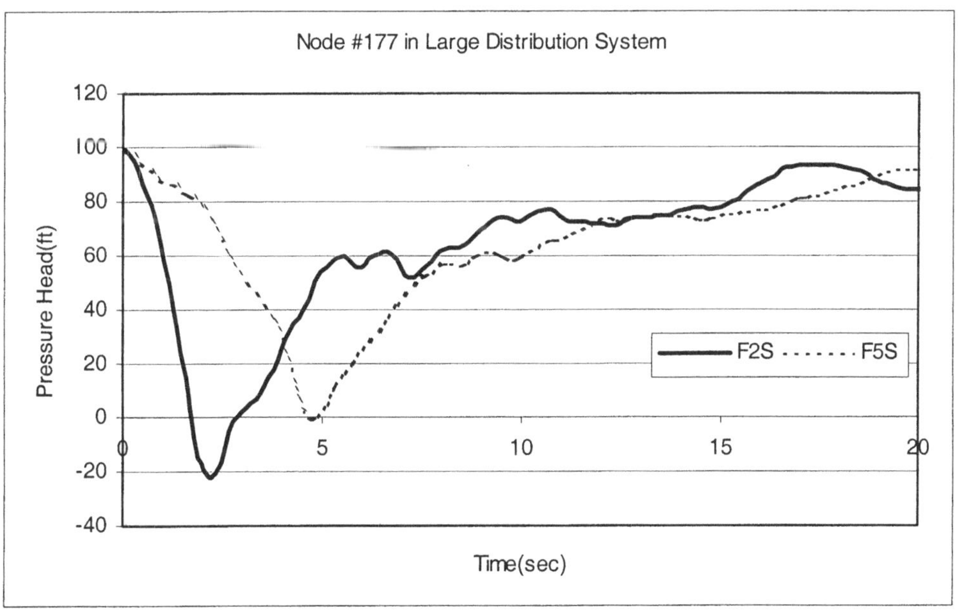

Figure 4.45 Comparison of pressure head at Node #177 for different hydrant opening rates (Model City A, 2 sec vs. 5sec)

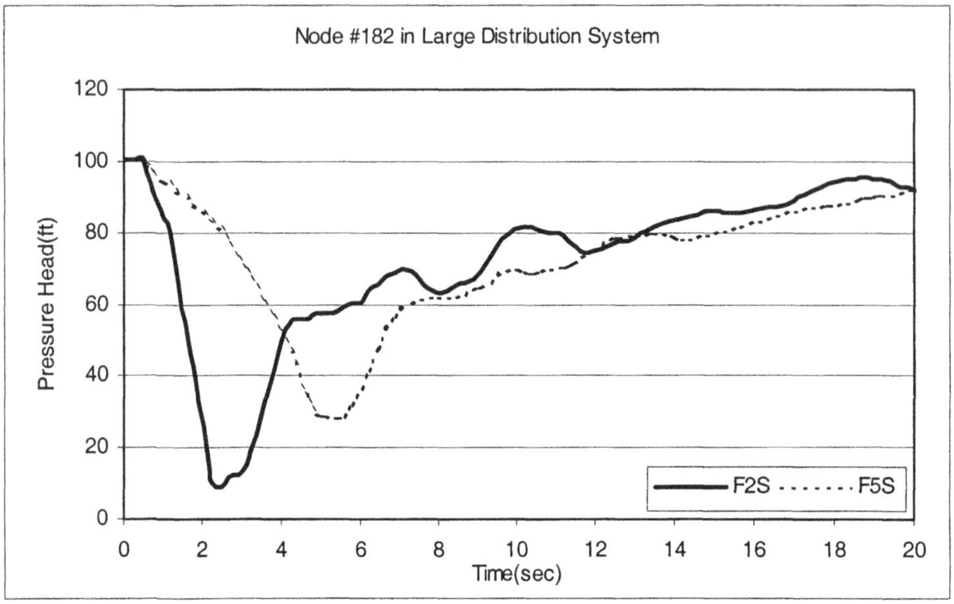

Figure 4.46 Comparison of pressure head at Node #182 for different hydrant opening rates (Model City A, 2 sec vs. 5 sec)

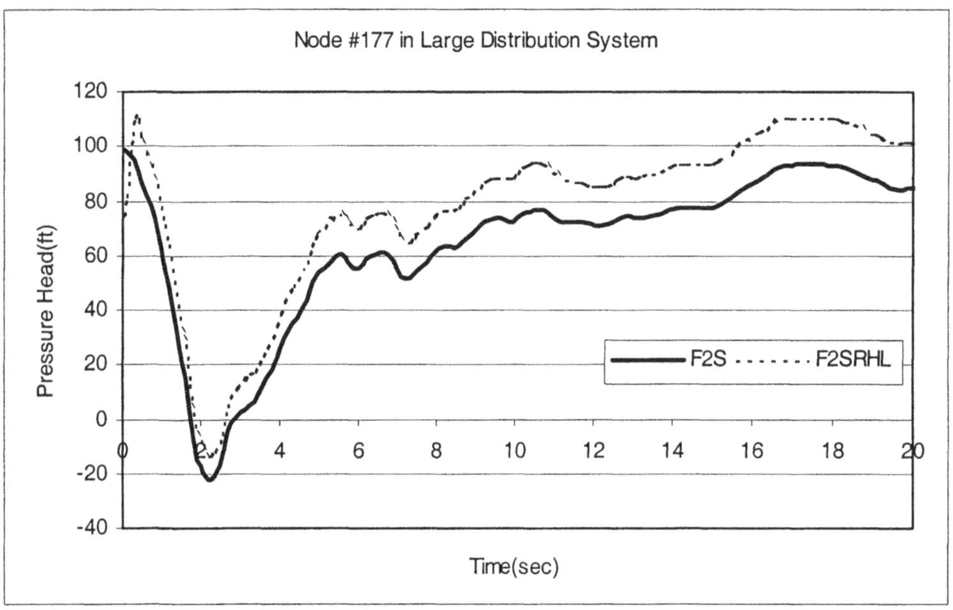

Figure 4.47 Comparison of pressure head at Node #177 for different reservoir water levels (Model City A)

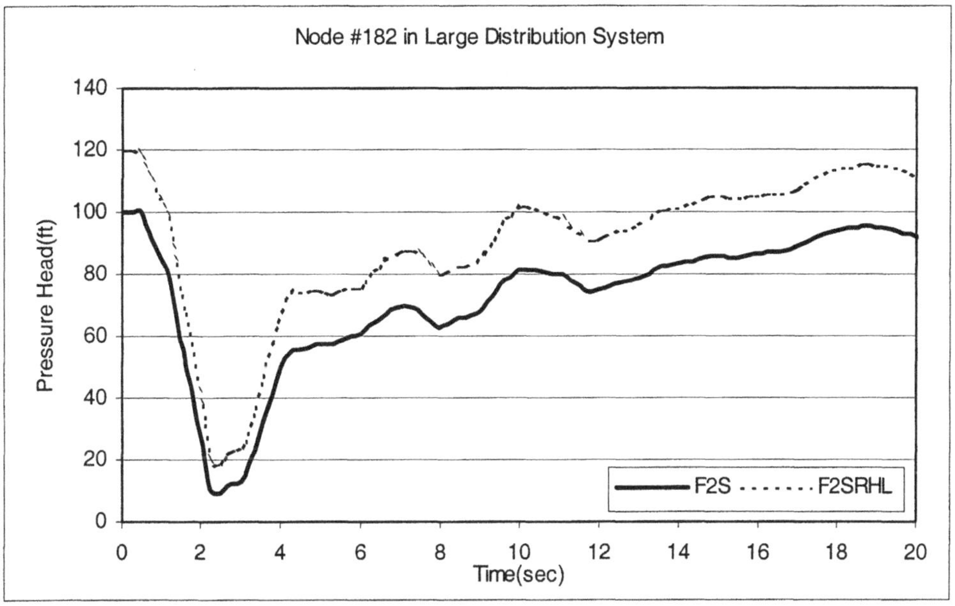

Figure 4.48 Comparison of pressure head at Node #182 for different reservoir water levels (Model City A)

CHAPTER 5
CONCLUSIONS AND RECOMMENDATIONS

SUMMARY AND CONCLUSION

Major conclusions from the present study are summarized in the following:

1. No geographic trends were found in North America in relation to the type of cross-connections being experienced. Indirect cross-connections account for 65% of the cross-connections while 35% are direct cross-connections.

2. Cross-connections that were bacteriological in nature accounted for 28.8% of the cross-connections, 26.1% were recorded as chemical in nature and 29.8%, a combination of both. "Other" was indicated for 15.6% of the cross-connections.

3. 38% of the respondents indicated they did indeed experience backflow incidents while 42% indicated having not experienced an incident and 19% indicated "unknown".

4. The North American survey data indicates that more and more agencies developed programs over the last three decades. As more agencies developed programs, more incidents were reported. Therefore, the reporting of incidents has continually increased since 1975. It is important, however, to note that the average number of incidents being reported per agency does not significantly change. This suggests that the number of incidents per utility is not necessarily changing over time, but that reporting is becoming more prevalent as agencies develop cross-connection control programs.

5. Agencies with cross-connection control programs report incidents more readily than agencies without a program. Agencies without a program are less likely to be aware of backflow incidents and therefore, not as likely to report them.

6. The agencies indicated that 83% of documented cross-connections were inside the premise of the customer, while 16% were in areas under the direct control of the water distribution system.

7. The overall average for all sizes of water utilities is $44,835 spent annually on administering the cross-connection control program. On the average 3.65% of the agencies' operations and maintenance budget was directed towards cross-connection control.

8. Reporting agencies on the average indicated that, per incident, the cost to the agency was approximately $4,201 in direct costs. $1,292 in indirect costs and $10,650 in personnel costs for an overall average cost of $16,143 per incident. It is important to note that the range of expenses associated with a backflow incident was dramatic. Some agencies had figures as low as $25, $7, or $0, while some had expenses of over $1,000,000 with one even indicating $1.6 million.

9. The data analysis showed that the nominal working pressure in a water system does have an effect on the number of backflow incidents experienced. Systems with a nominal working water pressure of more than 100 PSI, experienced less backflow incidents than those systems with lower nominal water pressures. Larger utilities experienced more backflow incidents.

10. The data analysis demonstrated that those utilities with higher maximum elevations differences experienced a greater number of backflow incidents. Analyzing a combination of nominal system pressure and maximum elevation difference indicated that systems with lower nominal pressure and higher elevation differences experienced more incidents than systems with higher nominal pressure and lower elevation differences. This conclusion is also demonstrated in the computer modeling.

11. Ninety-one percent of the utilities responding to the North American survey indicated that they did have a cross-connection control program. The survey data shows that 43% of the utilities surveyed administered system protection programs (their goal is to protect the water system at the users water connection).

12. A requirement to field test backflow preventers was indicated by 89% of the utilities responding, indicating that almost all utilities requiring a specific approval or listing of backflow prevention assemblies also has a requirement to field test the assemblies. A requirement to keep records of the backflow preventer field tests was indicated by 84%. Site surveys were required by 80% of the utilities to determine the extent of cross-connections on a user's property. However, only 57% had a means to determine who was qualified to conduct such site surveys. When a backflow incident does occur, the survey indicates only 49% of the utilities require that the incident be reported.

13. Establishing a legal authority was ranked as the most effective element of a cross-connection control program. Having a list of approved backflow preventers and then a requirement to test the backflow preventers followed this.

14. Fifty-three percent of the respondents indicate using termination of service as a means of gaining compliance, while only 44% indicate this as the most effective means of gaining said compliance. Respondents indicated that 60% of their customers believe that a cross-connection control program is not needed; specifically indicating that there is no perceived risk.

15. Many experts involved in cross-connection control believe that public education is the most effective means of bringing customers into compliance. However, only 15% of the survey respondents currently ranked public education as the most effective method of enforcement.

16. Results from the computer modeling clearly indicate that sudden changes in the water demand (or hydrant opening) caused drastic transient flow conditions resulting in

significant pressure loss and flow reversal in a network system of any size. Many examples demonstrate that the loss in pressure could reach below atmospheric pressure.

17. Distribution systems that have greater variance in ground elevations will have greater possibility of creating sub atmospheric pressure when drastic pressure reduction occurs.

18. Slower hydrant opening rates significantly reduce the impact of hydraulic transients created.

19. The extent of the affected area cannot be generalized. The affected area depends on the specific condition creating sudden changes and the particular system characteristics of the network.

RECOMMENDATIONS TO UTILITIES

An Effective Cross-Connection Control Program

Recommendations from the Surveys

The results of the project show that a cross-connection control program is necessary for water utilities from both the standpoint of public health and the potential financial impact of backflow incidents. The survey results show there are several elements that should be included in an effective cross-connection control program. Those surveyed indicated a legal authority (some sort of rule, regulation or ordinance) governing cross-connection control was the most effective element of the cross-connection control program. A list of acceptable backflow preventers along with a requirement to field-test these backflow preventers were also listed as very effective elements of the program. Record keeping also ranked fairly high as a required element of a cross-connection control program. There were some other elements of a program that were not listed as one of the most effective elements. A further study of the survey results, however, shows that some of these elements should be considered necessary in any cross-connection control program. One of these is the requirement to report backflow incidents. Although, this was not listed as a requirement by most utilities, it was evident that there is a lack of documentation of backflow incidents.

Another point of major concern, which came up several times in the survey, was the necessity for public education. Although, public education was not listed as one of the most effective elements of a program, uninformed public and uninformed water utility personnel appeared to be a major problem in the administration of a cross-connection control program. Since the public and even utility personnel are not knowledgeable in the area of cross-connection control public education should be considered a high priority.

In summary, water utilities should initiate and/or maintain cross-connection control programs, which include:

1. A legal document governing the program
2. The use of listed or approved backflow prevention assemblies
3. Field-testing the backflow prevention assemblies
4. Trained, licensed, or certified personnel to field-test the backflow preventers

5. Site surveys to determine the extent of backflow prevention required
6. A requirement to keep records and document backflow incidents
7. A public education effort to both the general public and water utility personnel

Recommendations from the Computer Modeling

The computer modeling provided clear evidence that sudden change in demand can cause drastic pressure drops and flow reversals. This may be due to main breaks, fire hydrant use, or other changes in water demand. The impact of rapid pressure drop and flow reversal could be lessened if the rate of change is decreased. Although this may not be possible in the case of a main break, it is possible in cases of flushing hydrants and other controlled changes in the system. Operations personnel need to be trained to make these changes gradually in order to lessen the effect of rapid system changes. The computer modeling, along with the survey results, showed that systems with larger elevation differences are more susceptible to the negative impact of sudden changes of water demand in the system.

Since the evidence shows that sudden changes in demand can clearly create undesirable conditions within the distribution system, appropriate plans to prevent unwanted flow reversals from occurring should be made prior to potential disasters. These plans can only be made after a careful study of the response of the system to various assumed scenarios.

APPENDIX A
INITIAL SURVEY

USC

UNIVERSITY OF SOUTHERN CALIFORNIA

Foundation for Cross-Connection Control and Hydraulic Research

National Assessment of the Impact of Cross Connection s in North American Water Supplies

American Water Works Association Research Foundation
Project 2611

Please check the information below and make any corrections necessary.

Name____

Company Name________________________________

Address______________________________________

City, State Zip_________________________________

General Information

1. Please classify your water agency

(1)[ ] CWS (Community Water System)
(2)[ ] NTNC (Non-Transient, Non-Community)
(3)[ ] TNC (Transient, Non-Community)

2.. *Please classify your water agency*

(1)[ ] Publicly owned
(2)[ ] Privately owned
(3)[ ] Don't know

Hydraulics of your System

3. Which of the following has your agency experienced in the last year? Check all that apply.

(1)[ ] Low or fluctuating pressure in the water system
(2)[ ] Main breaks or leaks due to aging infrastructure
(3)[ ] Main breaks or leaks due to human error
(4)[ ] Inadequate distribution storage (demand greater than supply)
(5)[ ] Improper sized piping due to system growth
(6)[ ] High demand (fire fighting, system flushing, etc.)
(7)[ ] Excessive elevation differences
(8)[ ] Flushing program activity

4. How many main breaks per mile of system piping per year does your system experience?

________________ breaks/mile/year

5. What is the nominal working pressure in your distribution system?

(1)[ ]	20-40 psi	(138-276 Kpa)
(2)[ ]	41-60 psi	(277-414 Kpa)
(3)[ ]	61-80 psi	(415-552 Kpa)
(4)[ ]	81-100 psi	(553-689 Kpa)
(5)[ ]	101-125 psi	(690-861 Kpa)
(6)[ ]	Greater than 125 psi	(Greater than 861 Kpa)

6. What is the nominal pressure at the most remote point of your distribution system?

(1)[ ]	20-40 psi	(138-276 Kpa)
(2)[ ]	41-60 psi	(277-414 Kpa)
(3)[ ]	61-80 psi	(415-552 Kpa)
(4)[ ]	81-100 psi	(553-689 Kpa)
(5)[ ]	101-125 psi	(690-861 Kpa)
(6)[ ]	Greater than 125 psi	(Greater than 861 Kpa)

7. *What is the maximum difference in elevation in your distribution system? For example if your distribution system operates at sea level and also at a maximum elevation of 300 feet above sea level, your response would be "251 to 500 feet."*

(1)[ ] Less than 50 feet (Less than 15.2 meters)
(2)[ ] 50 to 100 feet (15.3 to 30.5 meters)
(3)[ ] 101 to 250 feet (30.6 to 76.2 meters)
(4)[ ] 251 to 500 feet (76.3 to152.4 meters)
(5)[ ] Greater than 500 feet (Greater than 152.4 meters)

8. *What is your typical pressure fluctuation in your system throughout the day. For example if your highest pressure is 100 psi and your system drops to 85 psi during peak demands, your answer would be "10 to 20 psi."*

(1)[ ] Less than 5 psi (Less than 35 Kpa)
(2)[ ] 5 to 10 psi (36 to 69 Kpa)
(3)[ ] 11 to 20 psi (70 to 138 Kpa)
(4)[ ] 21 to 30 psi (139 to 207 Kpa)
(5)[ ] 31 to 40 (208 to 276 Kpa)
(6)[ ] Greater than 40 psi (Greater than 276 Kpa)

9. *How many water connections are there in your system?*

(1)[ ] Less than 500
(2)[ ] 500 to 1000
(3)[ ] 1001 to 5000
(4)[ ] 5001 to 25,000
(5)[ ] 25,001 to 100,000
(6)[ ] Greater than 100,000

10. *How many water connections are there in your system in each of the following categories?*

(1)__________Commercial/Industrial Domestic
(2)__________Residential Domestic
(3)__________Irrigation
(4)__________Fire suppression

11. *What is the daily water usage in your distribution system?*

(1)[ ] less than 10,000 gallons/day (Less than 37,900 liters/day)
(2)[ ] 10,001 to 100,000 gallons/day (37,901 to 379,000 l/day)
(3)[ ] 100,001 to 1,000,000 gallons/day (379,001 to 3,790,000 l/day)
(4)[ ] 1,000,001 to 10,000,000 g/day (3,790,001 to 37,900,000 l/day)
(5)[ ] 10,000,001 to 50,000,000 g/day (37,900,001 to 189,500,000 l/day)
(6)[ ] Greater than 50,000,000 g/day (Greater than 189,500,000 l/day)

The Cross-Connection Control Program

12. *Does your agency currently have a cross-connection control program?*

(1)[ ] Yes
(2)[ ] No

13. *Which of the following sources or recommended practices do you utilize?*

(1)[ ] AWWA Manual M-14
(2)[ ] USC Manual of Cross-Connection Control, Ninth Edition
(3)[ ] EPA Cross-Connection Control Manual.
(4)[ ] Other ____________________

14. *What elements are included in your cross-connection control program? (Check all that apply.)*

(1)[ ] Legal authority (Some legal document {i.e., ordinance, rule, code, etc.} giving the agency the authority to carry out a cross-connection control program)
(2)[ ] Acceptable backflow preventers (A listing/approval of specific backflow prevention assemblies which are acceptable under the agency's jurisdiction)
(3)[ ] A requirement to field test backflow preventers
(4)[ ] An acceptable method of field testing backflow prevention assemblies
(5)[ ] A means to determine who is qualified to field test backflow prevention assemblies (This may be the agency's certification/licensing program, or the adoption of a third party tester certification program)
(6)[ ] Conducting site surveys to determine the extent of cross-connections on a premise.
(7)[ ] A means to determine who is qualified to conduct site surveys (This may be the agency's certification/licensing program, or the adoption of a third party survey/inspection certification program)
(8)[ ] A requirement to keep records of backflow preventer field test results
(9)[ ] A public education effort to educate the public in the area of cross-connection control
(10)[ ] Requirement for Reporting of backflow incidents

15. *Please rank the following in order of effectiveness (1 being the most effective). If you do not have the element, please leave the ranking blank.*

(1)____Legal authority (Some legal document {i.e., ordinance, rule, code, etc.} giving the agency the authority to carry out a cross-connection control program)

(2)___Acceptable backflow preventers (A listing/approval of specific backflow prevention assemblies, which are acceptable under the agency's jurisdiction)
(3)___A requirement to field test backflow preventers
(4)___An acceptable method of field-testing backflow prevention assemblies
(5)___A means to determine who is qualified to field test backflow prevention assemblies (This may be the agency's certification/licensing program, or the adoption of a third party tester certification program)
(6)___Conducting site surveys to determine the extent of cross-connections on a premise.
(7)___A means to determine who is qualified to conduct site surveys (This may be the agency's certification/licensing program, or the adoption of a third party survey/inspection certification program)
(8)___A requirement to keep records of backflow preventer field test results
(9)___A public education effort to educate the public in the area of cross-connection control
(10)__Requirement for Reporting of backflow incidents

16. Do you conduct site surveys to determine the extent of cross-connections on a premise as part of your cross-connection control program?
(1)[ ] Yes
(2)[ ] No

17. Does your agency have legal authority (Some legal document {i.e., ordinance, rule, code, etc.} giving the agency the authority to carry out a cross-connection control program) covering cross-connection control?
(1)[ ] Yes
(2)[ ] No

18. What are the most common cross-connections found during site surveys? (Check all that apply.)
(1)[ ] Irrigation
(2)[ ] Swimming Pools/Spas
(3)[ ] Fire Suppression Systems
(4)[ ] Carbonation Systems
(5)[ ] Cooling Towers
(6)[ ] Boilers
(7)[ ] Tanks (onsite storage tank or vats)?
(8)[ ] Food processing/preparationEquipment
(9)[ ] Garden/Washdown hoses
(10)[ ] Toilets, Urinals and other sewage connections
(11)[ ] Wells
(12)[ ] Other______________________________

19. At which types of facilities do you most commonly find cross-connections? (Check no more than three.)

(1)[ ] Medical
(e.g., hospitals, morgues, dental, nursing homes, clinics, etc.)
(2)[ ] Schools
(e.g., elementary, high schools, colleges, universities)
(3)[ ] Manufacturing
(e.g., automotive, chemical, paper, plating)
(4)[ ] Food Processing
(e.g., restaurants, dairies, beverage plants, canneries, etc.)
(5)[ ] Services (e.g., Laundries, car washes, film labs, etc.)
(6)[ ] Restricted (e.g., civil works, classified, other closed facilities)
(7)[ ] Sewage Treatment
(8)[ ] Residential
(9)[ ] Other________________

20. In site surveys what percentage of cross-connections found are direct vs. indirect cross-connections? (The total should add up to 100%.)

(1)____% Indirect cross-connections (those subject to backsiphonage only)
(2)____% Direct cross-connections (those subject to backpressure and backsiphonage.)

21. In site surveys what percentage of cross-connections found are non-health, health, or sewage hazards? (The total should add up to 100%.)

(1)____% Non-Health Hazards
(2)____% Health Hazards
(3)____% Sewage

22. In site surveys what percentage of cross-connections found are bacteriological, chemical, or physical in nature? (The total should add up to 100%.)

(1)____% Bacteriological
(2)____% Chemical
(3)____% a combination of (1) and (2)
(4)____% other (i.e., a physical problem such as compressed air)

23. What type of cross-connection control program does your agency have?

(1)[ ] System protection, or containment (backflow protection at the meter or service connection)
(2)[ ] Internal protection, or isolation (backflow protection at each point of water use)
(3)[ ] Combination of System and Internal Protection
(4)[ ] None

24. *If your agency uses System Protection, how many of each backflow preventer are installed in your system for system protection?*

(1)___Air Gap
(2)___Reduced pressure principle assembly
(3)___Reduced pressure principle detector assembly
(4)___Double check valve assembly
(5)___Double check detector assembly
(6)___Pressure vacuum breaker assembly/Spill-resistant vacuum breaker
(7)___Atmospheric vacuum breaker assembly
(8)___Other_______________
(9)___Not Applicable

25. *If your agency uses Internal Protection, how many of each backflow preventer are installed in your system for internal protection?*

(1)___Air Gap
(2)___Reduced pressure principle assembly
(3)___Reduced pressure principle detector assembly
(4)___Double check valve assembly
(5)___Double check detector assembly
(6)___Pressure vacuum breaker assembly/Spill-resistant vacuum breaker
(7)___Atmospheric vacuum breaker assembly
(8)___Other_______________
(9)___Not Applicable

26. *If your agency uses combined System/Internal Protection, how many of each backflow preventer are installed in your system?*

(1)___Air Gap
(2)___Reduced pressure principle assembly
(3)___Reduced pressure principle detector assembly
(4)___Double check valve assembly
(5)___Double check detector assembly
(6)___Pressure vacuum breaker assembly/Spill-resistant vacuum breaker
(7)___Atmospheric vacuum breaker assembly
(8)___Other_______________
(9)___ Not Applicable

27. *Which of the following are included in your legal document? Check all that apply*

(1)[ ] A requirement for site surveys to determine if backflow protection is necessary.
(2)[ ] A listing/approval requirement for acceptable backflow preventers
(3)[ ] A requirement for backflow preventers to be field tested periodically
(4)[ ] A specific method for field testing backflow preventers
(5)[ ] A specific certification/licensing of backflow prevention assembly testers

(6)[ ] That records of backflow assembly installations, field tests, and repairs be made and kept for a period of time
(7)[ ] Requirement for reporting backflow incidents

28. How often are backflow prevention assemblies under your agency's jurisdiction required to be field tested?
(1)[ ] Every three months
(2)[ ] Every six months
(3)[ ] Every twelve months
(4)[ ] Every twenty-four months
(5)[ ] Other_______________
(6)[ ] No requirement

29. What field test procedure is used to field test backflow preventers?
(1)[ ] USC Manual of Cross-Connection Control, Ninth Edition
(2)[ ] USC Manual of Cross-Connection Control, Eighth Edition
(3)[ ] ASSE Series 5000, First Edition
(4)[ ] New England Water Works Association procedures
(5)[ ] Other__________________________

30. What methods of enforcement have been used to bring customers into compliance with the requirements of the backflow prevention program? (Check all that apply)
(1)[ ] Compliance letters/written notices
(2)[ ] Fines
(3)[ ] Civil citations
(4)[ ] Criminal citations
(5)[ ] Termination of water service
(6)[ ] Public Education
(7)[ ] ___________________________________

31. What method of enforcement do you consider to be the most effective in bringing customers into compliance with the requirements of the backflow prevention program? (Check one)
(1)[ ] Written notices
(2)[ ] Fines
(3)[ ] Civil citations
(4)[ ] Criminal citations
(5)[ ] Termination of service
(6)[ ] Public Education
(7)[ ] Other___________________________________

32. Should corrective actions be necessary, (i.e., installation of backflow preventer, discontinuance of water service, etc.) what are the factors in determining a period for compliance for the existing systems?

(1)[ ] Case by case
(2)[ ] Degree of hazard presented in the system
(3)[ ] Other

33. Should corrective actions be necessary (i.e., installation of backflow preventer, discontinuance of water service, etc.), what is your typical timeframe for compliance?
(1)[ ] Immediate
(2)[ ] 1-10 days
(3)[ ] 11-30 days
(4)[ ] 31-60 days
(5)[ ] 61-90 days
(6)[ ] 91-120 days
(7)[ ] more than 120 days
(8)[ ] Other __________

34. What methods of public education have you used to educate people about the hazards of backflow? (check all that apply)
(1)[ ] News Media
(2)[ ] Cable TV
(3)[ ] Newspapers
(4)[ ] Public Events (e.g., AWWA Blue Thumb week)
(5)[ ] Bill Stuffers
(6)[ ] Other______________

33. Which was found to be the most effective?
(1)[ ] News Media
(2)[ ] Cable TV
(3)[ ] Newspapers
(4)[ ] Public Events (e.g., AWWA Blue Thumb week)
(5)[ ] Bill Stuffers
(6)[ ] Other ______________

34. How do you determine the effectiveness of the public education method?
(1)[ ] Less resistance to backflow requirements
(2)[ ] Fewer complaints about Cross-Connection Control Program
(3)[ ] Fewer backflow incidents
(4)[ ] Other__________________________

35. Does your cross-connection control program coordinate efforts with any of the following? Check all that apply.

(1)[ ] State Health Department
(2)[ ] Local Health Agency
(3)[ ] Other local water agencies
(4)[ ] Plumbing Code enforcement agency
(5)[ ] Fire department
(6)[ ] Other______________________________
(7)[ ] Organizations (such as AWWA, ABPA, etc.)

36. Which of the following customer complaints has your agency received as a result of conducting/administering your cross-connection control program? Check all that apply.

(1)[ ] CCC program is governmental intervention
(2)[ ] Unneeded (no perceived risk)
(3)[ ] Reduction in available water pressure from the installation of backflow prevention assembly
(4)[ ] Reduction in available water flow due to installation of backflow prevention assembly
(5)[ ] Cost (installation, testing, maintenance of backflow prevention assembly)
(6)[ ] Interruption of water service due to periodic field testing of backflow prevention assembly
(7)[ ] Other____________

37. What are the major limiting factors in implementing or administrating a cross-connection control program

(1)[ ] Costs involved
(2)[ ] Water agency not convinced the program is important
(3)[ ] Conflicts with other jurisdictions or administrative authorities
(4)[ ] Consumers uninformed
(5)[ ] Politics
(6)[ ] Other ____________

38. What is the estimated yearly cost of running your cross-connection control program?

Personnel $________
Equipment/Material $ ________
Other $__________

Backflow Incidents

39. Has your agency experienced a backflow incident?

(1)[ ] Yes
(2)[ ] No
(3)[ ] Unknown

40. How many incidents (documented or undocumented) has your agency experienced since 1970?

(1)[ ] 1-2
(2)[ ] 3-5
(3)[ ] 6-10
(4)[ ] 11-20
(5)[ ] 21-50
(6)[ ] 51-100
(7)[ ] more than 100

41. How many of these have been documented?

(1)[ ] 1-2
(2)[ ] 3-5
(3)[ ] 6-10
(4)[ ] 11-20
(5)[ ] 21-50
(6)[ ] 51-100
(7)[ ] more than 100

42. What percentage of cross-connections you have documented are internal cross-connections (meaning those on the water user's premises) versus cross-connections under the direct control of the distribution system (such as a fire hydrant to a hose connection, cross-connected water mains, etc).

(1)_____% internal cross-connections
(2)_____% distribution cross-connections

43. What were the physical effects of the incident(s)? (Check all that apply)

(1)[ ] None
(2)[ ] Water Quality Complaint by Customer(s)
(3)[ ] Public Notice Issued (Boil water order, or notice of water quality being compromised)
(4)[ ] Interruption of Water Service
(5)[ ] Damage to the Distribution System
(6)[ ] Replacement of Distribution components (i.e., pipes, valves, etc.)
(7)[ ] Illness(es)
(8)[ ] Death(s)

44. What financial effects of the incident(s) did the agency realize?

(1)[ ] None
(2)[ ] Investigation (on-site, administrative, etc.)
(3)[ ] Analysis (water sampling, etc.)
(4)[ ] Clean-up (flushing, decontamination, etc.)
(5)[ ] Damage to the Distribution System
(6)[ ] Replacement of Distribution components (i.e., pipes, valves, etc.)
(7)[ ] Lost revenue
(8)[ ] Backflow Prevention Assembly (Purchase and/or Installation)
(9)[ ] Other_______________
(10)[ ] Unknown

45. On the average per each incident what would you estimate the dollar amount cost was to the agency?

(1)$________ in Direct Costs (Replacing pipes, components, etc.)
(2)$________ in Indirect Costs (Lost revenue, etc.)
(3)$________ in Personnel Costs

46. What financial effects of the incident(s) did the consumer realize?

(1)[ ] None
(2)[ ] Lost Revenue (down time, equipment out of service, etc.)
(3)[ ] Personnel (sick time, workman's compensation, etc.)
(4)[ ] Clean-up (repair/replacement of affected equipment)
(5)[ ] Backflow Prevention Assembly (purchase and/or installation)
(6)[ ] Other_______________
(7)[ ] Unknown

47. On the average per each incident what would you estimate the dollar amount cost was to the consumer?

(1)$________ in Direct Costs (Replacing pipes, components, etc.)
(2)$________ in Indirect Costs (Lost revenue, etc.)
(3)$________ in Personnel Costs

48. In the space below, please describe some of the common problems you have experienced in the administration of your cross-connection control program.

____

____

____

APPENDIX B
FOLLOW-UP SURVEY

USC

UNIVERSITY
OF SOUTHERN
CALIFORNIA

AWWARF

Project 2611

National Assessment of the Impact of Cross-Connections in North American Water Supplies

Follow-up Survey

Please make any corrections necessary to this label.

1. What percentage of your service connections are metered?

______%

2. If not all of your water service connections are metered, which types of services are not metered? (Please check all that apply.)

____Residential
____Fire Protection
____Construction

____Other____

3. Do you have a means of determining if your water meters run backwards? (This information is likely available from your billing department.)

____Yes
____No

4. If so, please provide the number of meters that have been documented as running backwards over the past twelve months.

____

5. If the information is available, please provide the number of water meters, which have been documented as running backwards for each of the following years: (If the information is not available please indicate by marking "N/A."

1998______
1997______
1996______
1995______

6. How was this information documented?

____Billing information
____Other______________________.

7. Have you documented that fewer meters have been running backwards since you initiated a cross-connection control program?

____Yes
____No
____Don't Know
____The question does not apply

8. Does your system have a regular flushing program?

____Yes
____No

9. If your system has a regular flushing program, how often is the system flushed?
Every _____months.

10. Have you experienced any water quality complaints, problems or other water system operational problems as a result of flushing activities?
____Yes
____No

11. If yes what concerns were experienced?

____

12. What percentage of your service connections are (should add up to 100%)
_____ ½" connections
_____¾" connections
_____1" connections
_____1 ½" connections
_____2" connections
_____3" connections
_____4" connections
_____6" connections
_____8" connections
_____10" connections
_____12" connections
_____Other__________connections

13. In the initial questionnaire question 24 read, "If your agency uses system protection, how many of each backflow preventer are installed in your system for system protection?" The options were.

- Air gap
- Reduced pressure principle assemblies
- Reduced pressure principle detector assemblies
- Double check valve assemblies
- Double check detector assemblies
- Pressure vacuum breaker assemblies
- Atmospheric vacuum breaker assemblies
- Atmospheric vacuum breaker assemblies
- OTHERS
- Not applicable.

If you marked (or would mark) "Others," please specify what other types of backflow preventers are used.

____

____

14. In the initial questionnaire question 31 read, "What method of enforcement do you consider to be the most effective?" If you selected "Termination of service," could you please clarify?

____The water service is actually turned off
____The customer is notified that the water services will be turned off unless he/she complies with your requirements.
____Both of the above

15. Question 10 of the initial questionnaire stated, "How many water connections are there in your system in each of the following categories?" Could you also provide the number of these that currently have backflow protection at the service connection?

Total number of water connections:
(1)__________Commercial/Industrial Domestic
(2)__________Residential Domestic
(3)__________Irrigation
(4)__________Fire suppression

Number of those with backflow protection
(1)__________Commercial/Industrial Domestic
(2)__________Residential Domestic
(3)__________Irrigation
(4)__________Fire suppression

16. Has the AWWARF report entitled, "Impact of Wet Pipe Fire Systems on Drinking Water Quality" had any effect on the way your agency addresses backflow requirements for fire protection water service connections?

____Yes
____No (go to Question 18)
____Haven't seen the report (go to Question 19)

17. If yes, what effect or changes resulted due to the information contained in the report?

____

____

18. If no, please state the reasons you feel the recommendations of the report are not being followed by your agency.

____Agency policies already consistent with recommendations.
____Conflicts with local codes/regulations
____Fire protection systems not considered a hazard

____Other____
____

19. Number 16 in the original questionnaire asked if you conducted site surveys. If you answered "No" to Number 16 in the original questionnaire, what is being done in lieu of site surveys?

____Blanket program, we require backflow preventers on all (check all that apply)
    ____Commercial services
    ____Residential services
    ____Fire services
    ____Irrigation services
    ____Other services______________
____Site surveys are conducted by another administrative authority
____Site surveys are conducted by a third party for our agency
____Survey conducted by mail in questionnaire
____Survey conducted by phone
____Other____________________
____Nothing else is being done.

20. Question 35 in the original questionnaire asked, "Does your cross-connection control program coordinate efforts with any of the following?

- State Health Department
- Local Health Department
- Other local Water Agencies
- Plumbing Code enforcement agency
- Fire Department
- Other
- Organizations (AWWA, ABPA, etc.)

Specifically what type of coordinated effort exists?

____Joint site surveys
____Regulatory support
____Enforcement actions
____Other________________________________

21. Question 40 of the original questionnaire stated, "How many incidents has your agency experienced since 1970?" To get a better understanding could you please complete the following, by filling in the number of incidents that have occurred for each period of time indicated?

_______ from 1971 through 1975.
_______ from 1976 through 1980.
_______ from 1981 through 1985.
_______ from 1986 through 1990.
_______ from 1991 through 1995.
_______ from 1996 through 2000.

22. In what year was your cross-connection control program imple mented (Please provide the year only).
_______on paper (adopted legal document)
_______in practice.

23. What approval or listing requirement is there for backflow prevention assemblies in your area of jurisdiction? (If there is a combination of these, please indicate all that apply to your area.)
____University of Southern California
____American Society of Sanitary Engineering
____International Association of Plumbing and Mechanical Officials
____Canadian Standards Association
____Underwriters Laboratory
____Factory Mutual
____Other_______________________________

24. How does your agency provide water? (Mark all that apply.)
____as a wholesaler
____as a retailer
____ to agricultural systems
____Other.

25. What is the average cost in your area of jurisdiction for each of the following for each size of ***reduced pressure principle assembly*** indicated:

	1"	4"	10"
Testing an assembly?	$_________	$_________	$_________
Repair of assembly?	$_________	$_________	$_________
Installation?	$_________	$_________	$_________
Purchase?	$_________	$_________	$_________

26. For field testing backflow preventers which does your agency use?

____In-house testers (i.e., employed by administrative authority)

____Private testers (i.e., commercial testers)

27. Does your agency operate a recycled water system?

____Yes

____No (go to Question 32)

28. If yes, in Question 27,do the recycled water services require back flow protection?

____Yes

____No (go to Question 32)

29. If yes, in question 28, what type of backflow protection is required?

____Double check valve assembly

____Reduced pressure principle assembly

____Pressure Vacuum Breaker Assembly

____Other____

30. If your agency operates a recycled system, do the potable water services to sites having recycled water require backflow protection?

____Yes

____No (go to Question 32)

31. If yes, in Question 30, what type of protection?

____Double check valve assembly

____Reduced pressure principle assembly

____Pressure Vacuum Breaker Assembly

____Other____

32. If you indicated in the first questionnaire that you have documented backflow incidents would you be willing to provide copies of the documentation?

____Yes (documentation is included with this questionnaire)

____Yes (documentation will be sent under separate cover)

____No, documentation will not be provided.

____There is no documentation of backflow incidents

33. How many water quality complaints has your agency received over the past ten years?

_____2000
_____1999
_____1998
_____1997
_____1996
_____1995
_____1994
_____1993
_____1992
_____1991

34. If you have a cross-connection control program, do you feel the program has had an effect on water quality complaints and problems?

____Yes
____No (go to Question 36)
____Not sure

35. If yes, please describe the effect your program has had on these complaints.

____

____

36. What percentage of your agency's annual Operations & Maintenance budget is directed towards cross connection control?

_____%

37. Does your agency have a Membership in or financially support any of the following organizations (check all that apply)?

____AWWA (American Water Works Association)
____USC (University of Southern California Foundation for Cross-Connection Control and Hydraulic Research)
____AWWARF (American Water Works Association Research Foundation)
____ABPA (American Backflow Prevention Association)
____Other______________________________

APPENDIX C
COMPLETE SUMMARY RESULTS OF THE FIRST SURVEY

No	Questions	Answers	Response	Percentage
-706	1. Please classify your water agency			
		1) CWS (Community Water System)	699	99%
		2) NTNC (Non-Transient, Non-Community)	4	1%
		3) TNC (Transient, Non-Community)	4	1%
-709	2. Please classify your water agency			
		1) Publicly owned	640	90%
		2) Privately owned	68	10%
		3) Don't know	1	0%
-682	3. Which of the following has your agency experienced in the last year? Check all that apply.			
		1) Low or fluctuating pressure in the water system	234	34%
		2) Main breaks or leaks due to aging infrastructure	527	77%
		3) Main breaks or leaks due to human error	252	37%
		4) Inadequate distribution storage (demand greater than supply)	106	16%
		5) Improper sized piping due to system growth	204	30%
		6) High demand (fire fighting, system flushing, etc.)	191	28%
		7) Excessive elevation differences	132	19%
		8) Flushing program activity	437	64%

-508	4. How many main breaks per mile of system piping per year does your system experience?	1) ________________ breaks/mile/year	5.7	0%
-710	5. What is the nominal working pressure in your distribution system?	1) 20-40 psi (138-276 Kpa)	41	6%
		2) 41-60 psi (277-414 Kpa)	256	36%
		3) 61-80 psi (415-552 Kpa)	259	36%
		4) 81-100 psi (553-689 Kpa)	104	15%
		5) 101-125 psi (690-861 Kpa)	33	5%
		6) Greater than 125 psi (Greater than 861 Kpa)	18	3%
-716	6. What is the nominal pressure at the most remote point of your distribution system?	1) 20-40 psi (138-276 Kpa)	197	28%
		2) 41-60 psi (277-414 Kpa)	300	42%
		3) 61-80 psi (415-552 Kpa)	112	16%
		4) 81-100 psi (553-689 Kpa)	48	7%
		5) 101-125 psi (690-861 Kpa)	35	5%
		6) Greater than 125 psi (Greater than 861 Kpa)	24	3%

-688	7. What is the maximum difference in elevation in your distribution system? For example if your distribution system operates			
		1) Less than 50 feet (Less than 15.2 meters)	121	18%
		2) 50 to 100 feet (15.3 to 30.5 meters)	112	16%
		3) 101 to 250 feet (30.6 to 76.2 meters)	173	25%
		4) 251 to 500 feet (76.3 to152.4 meters)	168	24%
		5) Greater than 500 feet (Greater than 152.4 meters)	114	17%
-708	8. What is your typical pressure fluctuation in your system throughout the day. For example if your highest pressure is 100			
		1) Less than 5 psi (Less than 35 Kpa)	114	16%
		2) 5 to 10 psi (36 to 69 Kpa)	305	43%
		3) 11 to 20 psi (70 to 138 Kpa)	223	31%
		4) 21 to 30 psi (139 to 207 Kpa)	38	5%
		5) 31 to 40 (208 to 276 Kpa)	17	2%
		6) Greater than 40 psi (Greater than 276 Kpa)	12	2%
-713	9. How many water connections are there in your system?			
		1) Less than 500	56	8%
		2) 500 to 1000	78	11%
		3) 1001 to 5000	258	36%

	4) 5001 to 25,000	217	30%
	5) 25,001 to 100,000	78	11%
	6) Greater than 100,000	27	4%
-636	10. How many water connections are there in your system in each of the following categories?		
	1) __________Commercial/Industrial Domestic	1983	0%
	2) __________Residential Domestic	19321	0%
	3) __________Irrigation	368	0%
	4) __________Fire suppression	671	0%
-705	11. What is the daily water usage in your distribution system?		
	1) less than 10,000 gallons/day (Less than 37,900 liters/day)	7	1%
	2) 10,001 to 100,000 gallons/day (37,901 to 379,000 l/day)	41	6%
	3) 100,001 to 1,000,000 gallons/day (379,001 to 3,790,000 l/day)	203	29%
	4) 1,000,001 to 10,000,000 g/day (3,790,001 to 37,900,000 l/day)	290	41%
	5) 10,000,001 to 50,000,000 g/day (37,900,001 to 189,500,000 l/day)	121	17%

	6) Greater than 50,000,000 g/day (Greater than 189,500,000 l/day)	42	6%
-704	12. Does your agency currently have a cross-connection control program?		
	1) Yes	639	91%
	2) No	66	9%
-622	13. Which of the following sources or recommended practices do you utilize?		
	1) AWWA Manual M-14	304	49%
	2) USC Manual of Cross-Connection Control, Ninth Edition	348	56%
	3) EPA Cross-Connection Control Manual.	171	27%
	4) Other ____________________	156	25%
-656	14. What elements are included in your cross-connection control program? (Check all that apply.)		
	1) Legal authority (Some legal document {i.e., ordinance, rule, code, etc.} giving the agency the authority to carry out a cr	610	93%
	2) Acceptable backflow preventers (A listing/approval of specific backflow prevention assemblies which are acceptable under t	588	90%
	3) A requirement to field test backflow preventers	581	89%
	4) An acceptable method of field testing backflow prevention assemblies	457	70%
	5) A means to determine who is qualified to field test backflow prevention assemblies (This may be the agency's certificatio	532	81%

6) Conducting site surveys to determine the extent of cross-connections on a premise.

522 80%

7) A means to determine who is qualified to conduct site surveys (This may be the agency's certification/licensing program, o

376 57%

8) A requirement to keep records of backflow preventer field test results

554 84%

9) A public education effort to educate the public in the area of cross-connection control

278 42%

10) Requirement for Reporting of backflow incidents

320 49%

15. Please rank the following in order of effectiveness (1 being the most effective). If you do not have the
-576 element, please

1) ___Legal authority (Some legal document {i.e., ordinance, rule, code, etc.} giving the agency the authority to carry out a

1.6 0%

2) ___Acceptable backflow preventers (A listing/approval of specific backflow prevention assemblies, which are acceptable und

3.4 0%

3) ___A requirement to field test backflow preventers

3.0 0%

4) ___An acceptable method of field-testing backflow prevention assemblies

4.6 0%

5) ___A means to determine who is qualified to field test backflow prevention assemblies (This may be the agency's certifica

		6) ___Conducting site surveys to determine the extent of cross-connections on a premise.	4.2	0%
		7) ___A means to determine who is qualified to conduct site surveys (This may be the agency's certification/licensing program	3.7	0%
		8) ___A requirement to keep records of backflow preventer field test results	5.6	0%
		9) ___A public education effort to educate the public in the area of cross-connection control	5.1	0%
		10) __Requirement for Reporting of backflow incidents	6.3	0%
-662	16. Do you conduct site surveys to determine the extent of cross-connections on a premise as part of your cross-connection co		6.7	0%
		1) Yes	530	80%
		2) No	132	20%

-674	17. Does your agency have legal authority (Some legal document {i.e., ordinance, rule, code, etc.} giving the agency the auth			
		1) Yes	613	91%
		2) No	62	9%
-610	18. What are the most common cross-connections found during site surveys? (Check all that apply.)			
		1) Irrigation	430	70%
		2) Swimming Pools/Spas	196	32%
		3) Fire Suppression Systems	295	48%
		4) Carbonation Systems	243	40%
		5) Cooling Towers	167	27%
		6) Boilers	277	45%
		7) Tanks (onsite storage tank or vats)?	148	24%
		8) Food processing/preparationEquipment	184	30%
		9) Garden/Washdown hoses	316	52%
		10) Toilets, Urinals and other sewage connections	126	21%
		11) Wells	203	33%
		12) Other________________________________	64	10%

-597	19. At which types of facilities do you most commonly find cross-connections? (Check no more than three.)			
		1) medical (e.g., hospitals, morgues, dental, nursing homes, clinics, etc.)	229	38%
		2) schools (e.g., elementary, high schools, colleges, universities)	194	32%
		3) Manufacturing (e.g., automotive, chemical, paper, plating)	219	37%
		4) Food Processing (e.g., restaurants, dairies, beverage plants, canneries, etc.)	220	37%
		5) Services (e.g., Laundries, car washes, film labs, etc.)	221	37%
		6) Restricted (e.g., civil works, classified, other closed facilities)	21	4%
		7) Sewage Treatment	43	7%
		8) Residential	216	36%
		9) Other_______________________________	55	9%
-520	20. In site surveys what percentage of cross-connections found are direct vs. indirect cross-connections? (The total should			
		1) ____% Indirect cross-connections (those subject to backsiphonage only)	65.8	0%
		2) ____% Direct cross-connections (those subject to backpressure and backsiphonage.)	35.0	0%

-513 21. In site surveys what percentage of cross-connections found are non-health, health, or sewage hazards? (The total should

1) ____% Non-Health Hazards

57.9 0%

2) ____% Health Hazards

41.0 0%

3) ____% Sewage

5.7 0%

22. In site surveys what percentage of cross-connections found are bacteriological, chemical, or physical
-451 in nature? (The to

1) ____% Bacteriological

29.1 0%

2) ____% Chemical

26.6 0%

3) ____% a combination of (1) and (2)

29.8 0%

4) ____% other (i.e., a physical problem such as compressed air)

16.2 0%

23. What type of cross-connection control program does your agency
-660 have?

1) System protection, or containment (backflow protection at the meter or service connection)

281 43%

	2) Internal protection, or isolation (backflow protection at each point of water use)	42	6%
	3) Combination of System and Internal Protection	313	47%
	4) None	24	4%
-239 24. If your agency uses System Protection, how many of each backflow preventer are installed in your system for system protec			
	1) ___Air Gap	7.2	0%
	2) ___Reduced pressure principle assembly	422.5	0%
	3) ___Reduced pressure principle detector assembly	16.0	0%
	4) ___Double check valve assembly	270.3	0%
	5) ___Double check detector assembly	103.9	0%
	6) ___Pressure vacuum breaker assembly	46.4	0%
	7) ___Spill-resistant vacuum breaker	0.2	0%
	8) ___Atmospheric vacuum breaker assembly		

6.9 0%

9) ___Other___________________

17.1 0%

10) ___Not Applicable

0.0 0%

-89 25. If your agency uses Internal Protection, how many of each backflow preventer are installed in your system for internal pr

1) ___Air Gap

29.5 0%

2) ___Reduced pressure principle assembly

114 0%

3) ___Reduced pressure principle detector assembly

113.3 0%

4) ___Double check valve assembly

163.9 0%

5) ___Double check detector assembly

4.3 0%

6) ___Pressure vacuum breaker assembly/Spill-resistant vacuum breaker

38.5 0%

7) ___Atmospheric vacuum breaker assembly

		0.5	0%
	8) ___Other__________________		
		26.4	0%
	9) ___Not Applicable		
		0.1	0%
-224	26. If your agency uses combined System/Internal Protection, how many of each backflow preventer are installed in your system		
	1) ___Air Gap		
		1093.6	0%
	2) ___Reduced pressure principle assembly		
		419.7	0%
	3) ___Reduced pressure principle detector assembly		
		5.6	0%
	4) ___Double check valve assembly		
		278.7	0%
	5) ___Double check detector assembly		
		71.8	0%
	6) ___Pressure vacuum breaker assembly/Spill-resistant vacuum breaker		
		187.0	0%
	7) ___Atmospheric vacuum breaker assembly		

			41.4	0%
		8) ___Other__________________	0.9	0%
		9) ___ Not Applicable	0.1	0%
-582	27. Which of the following are included in your legal document? Check all that apply			
		1) A requirement for site surveys to determine if backflow protection is necessary.	460	79%
		2) A listing/approval requirement for acceptable backflow preventers	469	81%
		3) A requirement for backflow preventers to be field tested periodically	538	92%
		4) A specific method for field testing backflow preventers	293	50%
		5) A specific certification/licensing of backflow prevention assembly testers	456	78%
		6) That records of backflow assembly installations, field tests, and repairs be made and kept for a period of time	461	79%
		7) Requirement for reporting backflow incidents	249	43%
-641	28. How often are backflow prevention assemblies under your agency's jurisdiction required to be field tested?			
		1) Every three months	3	0%
		2) Every six months	39	6%

	3) Every twelve months	540	84%
	4) Every twenty-four months	22	3%
	5) Other_______________	27	4%
	6) No requirement	53	8%
-489 29. What field test procedure is used to field test backflow preventers?			
	1) USC Manual of Cross-Connection Control, Ninth Edition	305	62%
	2) USC Manual of Cross-Connection Control, Eighth Edition	57	12%
	3) ASSE Series 5000, First Edition	43	9%
	4) New England Water Works Association procedures	37	8%
	5) Other__________________________	94	19%
-611 30. What methods of enforcement have been used to bring customers into compliance with the requirements of the backflow preve			
	1) Compliance letters/written notices	555	91%
	2) Fines	38	6%
	3) Civil citations	18	3%
	4) Criminal citations	8	1%
	5) Termination of water service	324	53%
	6) Public Education		

	Question	Response	Count	Percent
			233	38%
		7) ______________________________	32	5%
-630	31. What method of enforcement do you consider to be the most effective in bringing customers into compliance with the requir			
		1) Written notices	349	55%
		2) Fines	32	5%
		3) Civil citations	5	1%
		4) Criminal citations	1	0%
		5) Termination of service	282	45%
		6) Public Education	93	15%
		7) Other______________________________	45	7%
-637	32. Should corrective actions be necessary, (i.e., installation of backflow preventer, discontinuance of water service, etc.)			
		1) Case by case	370	58%
		2) Degree of hazard presented in the system	378	59%
		3) Other	34	5%
-611	33. Should corrective actions be necessary (i.e., installation of backflow preventer, discontinuance of water service, etc.),			
		1) Immediate	75	12%
		2) 1-10 days		

			69	11%
		3) 11-30 days	223	36%
		4) 31-60 days	152	25%
		5) 61-90 days	77	13%
		6) 91-120 days	19	3%
		7) more than 120 days	13	2%
		8) Other ___________	22	4%
-494	34. What methods of public education have you used to educate people about the hazards of backflow? (check all that apply)			
		1) News Media	42	9%
		2) Cable TV	29	6%
		3) Newspapers	109	22%
		4) Public Events (e.g., AWWA Blue Thumb week)	117	24%
		5) Bill Stuffers	265	54%
		6) Other________________	217	44%
-392	35. Which was found to be the most effective?			
		1) News Media	18	5%
		2) Cable TV	4	1%
		3) Newspapers	42	11%

4) Public Events (e.g., AWWA Blue Thumb week) 55 14%

5) Bill Stuffers 162 41%

6) Other ________________ 158 40%

36. How do you determine the effectiveness of the public education
-439 method?

1) Less resistance to backflow requirements 255 58%

2) Fewer complaints about Cross-Connection Control Program 144 33%

3) Fewer backflow incidents 71 16%

4) Other________________________ 44 10%

37. Does your cross-connection control program coordinate efforts with any of the following? Check all
-581 that apply.

1) State Health Department 357 61%

2) Local Health Agency 223 38%

3) Other local water agencies 109 19%

4) Plumbing Code enforcement agency 309 53%

5) Fire department 177 30%

6) Other___________________________ 68 12%

7) Organizations (such as AWWA, ABPA, etc.) 141 24%

			14	4%
-294	43. How many of these have been documented?			
		1) 1-2	193	66%
		2) 3-5	60	20%
		3) 6-10	24	8%
		4) 11-20	9	3%
		5) 21-50	3	1%
		6) 51-100	1	0%
		7) more than 100	4	1%
-379	44. What percentage of cross-connections you have documented are internal cross-connections (meaning those on the water user'			
		1) _____% internal cross-connections	83.0	0%
		2) _____% distribution cross-connections	16.2	0%
-450	45. What were the physical effects of the incident(s)? (Check all that apply)			
		1) None	226	50%
		2) Water Quality Complaint by Customer(s)	212	47%
		3) Public Notice Issued (Boil water order, or notice of water quality being compromised)	46	10%
		4) Interruption of Water Service		

			86	19%
		5) Damage to the Distribution System	24	5%
		6) Replacement of Distribution components (i.e., pipes, valves, etc.)	27	6%
		7) Illness(es)	22	5%
		8) Death(s)	1	0%
-447	46. What financial effects of the incident(s) did the agency realize?			
		1) None	233	52%
		2) Investigation (on-site, administrative, etc.)	171	38%
		3) Analysis (water sampling, etc.)	137	31%
		4) Clean-up (flushing, decontamination, etc.)	115	26%
		5) Damage to the Distribution System	15	3%
		6) Replacement of Distribution components (i.e., pipes, valves, etc.)	19	4%
		7) Lost revenue	31	7%
		8) Backflow Prevention Assembly (Purchase and/or Installation)	58	13%
		9) Other	11	2%
		10) Unknown	27	6%

	Question	Response	Count	Percent
-181	47. On the average per each incident what would you estimate the dollar amount cost was to the agency?	1) $ in Direct Costs (Replacing pipes, components, etc.)	4200.76	0%
		2) $ in Indirect Costs (Lost revenue, etc.)	1292.1	0%
		3) $ in Personnel Costs	10655.3	0%
-377	48. What financial effects of the incident(s) did the consumer realize?	1) None	146	39%
		2) Lost Revenue (down time, equipment out of service, etc.)	58	15%
		3) Personnel (sick time, workman's compensation, etc.)	19	5%
		4) Clean-up (repair/replacement of affected equipment)	86	23%
		5) Backflow Prevention Assembly (purchase and/or installation)	155	41%
		6) Other__________________________	17	5%
		7) Unknown	43	11%
-130	49. On the average per each incident what would you estimate the dollar amount cost was to the consumer?	1) $_________ in Direct Costs (Replacing pipes, components, etc.)	1422.7	0%

2) $__________ in Indirect Costs (Lost revenue, etc.)

1750	0%

3) $__________ in Personnel Costs

603.8	0%

APPENDIX D
COMMENTS ON COMMON PROBLEMS IN ADMINISTERING THE PROGRAM

- The lack of awareness the consumer has when discussing backflow issues has been our common problem.
- The need vs. the cost of a device (2" larger) plus the pressure loss differential (RP) at the point of connection is vital in keeping contaminates from entering the potable water system.
- The start up of a program for such a large utility. The record keeping has been the most difficult. Staffing problems and administration support.
- Political support.
- Customers believe this is a company responsibility and do not want to cooperate. They also believe the cost should be paid by the company, not through rates.
- Resentment of fees e.g. $75 for test, $40 for surveys.
- This is a new company just getting started.
- Support from management, discontinuance of water service for non-compliance, customer complaints.
- Our programs run pretty smoothly, there really haven't been any major problems.
- Problem getting the field test done by due date.
- Getting the approved testers to service the customer's device in a timely manner.
- Educating customers. Lack of sufficient time and resources to track the program. Public & political apathy.
- No personnel available due to cutbacks.
- We have just started the program. Our biggest problem has been training to allow us to inspect for problems and hazards.
- Politics hindering progress. Ordinance allows for termination of service but it's impossible to accomplish at large shopping centers. No programs in Fresno county health and city plumbing inspection.
- Once we received political and administrative support, it's been a lot easier.
- Biggest problem is being informed of new installations with or that should have devices. Another problem is getting customers to believe that devices are not needless government revenue.

- New program.
- Working with the customers to make the system compliant.
- Convincing owners of potential problems.
- People just don't know how precious safe and clean water is.
- No one takes cross-connection control seriously.
- Not enough time, money, personnel, cooperation.
- Money, support, politics.
- Lacking funds necessary to become compliant.
- Lack of training slots in backflow prevention.
- Educate the customer
- Water services in the vicinity of septic systems
- Compliance for installations and first year tests.
- Time constraints.
- Large effort to make sure all devices in the system are in the program. We continue to discover devices not in the program.
- People don't report on site violations.
- Uneducated public, state & federal personnel interfere with system operations
- The effectiveness of enrollment of compliance to the Cross-Connection Control program.
- Time and manpower.
- No active program and no knowledge of any problems.
- First contact, explain health risks, most understand after told.
- Politics and cost
- Going to gear up for a program soon.

- State and fed laws don't necessarily apply on a reservation
- We don't have a formal cc program. Under 15000 population. We do require cc devices on all new installations.
- We will be adapting a formal program in 2001.
- Public awareness.
- Manager sees no need for program and enforcement.
- You can't catch all cross-connections. You get what you can, but some slip through.
- Ignorance of architects & engineers. Regional problem here in TX.
- Residential garden hoses
- Time
- Resistance from contractors. Their resistance has led them to misinform their customers and therefore create bad PR for us.
- Need more personnel.
- Our program is new.
- We are forced to accept AVB to protect the water system. They can be installed and not inspected or tested annually. We believe them to be reliable but since USC accepts them, we must also. You should require annual inspections and because they are so easily modified.
- Timely testing, unqualified testers, false test reports, timely repair, misapplication of devices, incorrect review by building departments and fire agencies.
- Administrators believe above ground assemblies do not look good in the community. Private contractors do not like the municipality performing the test or repair, no retrofit program, 120+ unprotected fire lines.
- Resources to inspect all sites
- Program only deals with hydrants, fire sys. On site inspections. Do not happen yet. Program not developed and not ok'd by politicians.
- Convincing people it's necessary.

- Limited resources, lack of coop bet. Bldg. Dept and health dept.
- Finding time to carry out program. Bare bones current effort, little concern right now.
- Interpretation of the term potential for cross connections. Difference in plumb. Code vs. cross connect code.
- Spare parts
- Convincing upper management of the need. $$$$ in the way.
- Customers believe they're being picked on. Cost of installation.
- Eduucate consumer
- Resistance to backflow requirements, disbelief it can occur, cost of device, yearly test
- Difficult to get customers to comply
- Large amount of paperwork, public resistance to look of devices.
- Different interpretation of standards by muni & provincial plumb. Insp. Lack of training/awareness.
- Conflict between bldg. Code and plumb. Code, lack of by-law for enforcing backflow prevention, requirement for zone isolation US premise.
- Improper locations of backflow preventers, incorrect orientation of devices, wrong device installed for the hazard, poor penmanship of testers
- We do not have a problem that we know of, more time is needed to track these questions.
- Customers are lacking education
- No time and expertise to administer a program at this time. Town does have ordnance allowing for regulation.
- Relatively new system. Problem is customer. Wells connected to the public supply lines from the meter.
- Wholesaler of water to 5 municipalities.
- Education, cost, effective way to remedy
- Finding funds to retrofit systems. Manpower to test.

- Small city, small thinking, unimportant to administration
- Lack of survey experience, lack of manpower
- None - good public relations goes a long way
- Small system, no program.
- Small utility. Too time consuming.
- No backing by state Fire Marshall.
- Not enough time allowed to survey the system. Lack of support from management
- No state involvement
- CCC programs lead to unhappy customers which lead to unhappy voters which makes it hard to enforce.
- Time and manpower to test and repair, paperwork for documentation
- Dealing with irate customers
- Friction with costs involved.
- New program
- Not convinced of necessity.
- Customers don't understand. Retrofitting tough. State indifference.
- Customer delays in testing by due date. Improper installation of assemblies.
- Compliance issues, balking from local government, lack of pub education.
- Public education and goverment intervention
- Public uninformed
- None
- Political resistance. Too expensive.
- Resistance by customers

- Subjective inspections. No perfect recommended practice for everything.
- Non-compliance, public resistance, cost, politics.
- Political pressure
- Educating the public to the fact that the initial cost of the equipment outweighs the potential cost of an incident.
- Lack of state support
- Annual report from questions are confusing.
- Additional time to retest, keep current owner listing up to date.
- Goal is to have good historical data.
- Enthusiasm
- State legislation.
- Politics.
- Compliance.
- Manpower.
- Customer Complaints.
- Lack of enforcement.
- Time, personnel.
- Enforcement by state.
- Getting customers to do annual tests
- Educating the public why it's needed.
- Lack of time/personnel
- Convincing the customer that it can and does occur. Sloppy test reports.

- If we are supposed to be adding a cross-connection control program, nobody has brought it to my attention.
- Changes in policies and requirements of MA DEP
- Access to sites, cumbersome record keeping.
- Convincing customers of importance.
- Cost, politics, sell the consumer.

APPENDIX E
SUMMARY RESULTS OF THE FOLLOW-UP SURVEY

1. What percentage of your service connections are metered?
(342) 94.31%

2. If not all of your water service connections are metered, which types of services are not metered? (Please check all that apply.)
(233)

12.88% - Residential
63.52% - Fire Protection
11.59% - Construction
12.02% - Other

- seasonal
- municipal owned
- town facilities
- connections to cobb city water
- some unmetered fire services
- city
- hydrant connections
- commercial
- public
- some city facilities
- business
- some city buildings
- stock tanks
- some residential and some commercial
- irrigation
- city, parks
- irrigation, auto fire sprinklers
- district sites are unmetered
- some fire protec.
- parks
- detector lines metered, mains not
- city irrigation/bldgs
- fire fighting
- city owned facilities
- irrigation
- city services
- small median planters
- raw canal water

3. Do you have a means of determining if your water meters run backwards? (This information is likely available from your billing department.)
(345)

77.10% -Yes
22.90% - No

4. If so, please provide the number of meters that have been documented as running backwards over the past twelve months.
(68) 5.22

5. If the information is available, please provide the number of water meters, which have been documented as running backwards for each of the following years: (If the information is not available please indicate by marking "N/A."

(57)

1998: 6.04
1997: 2.19
1996: 2.2
1995: 2.5

6. How was this information documented?
(199)
84.92% - Billing information
15.08% - Other____________________.

1) service orders
2) inspection reports
3) meter reading
4) backflow prevention inspection report
5) meter reading
6) not documented
7) job codes in the meter repair group
8) field testing
9) meter operation
10) physical observation
11) zero usage
12) distribution employees & meter readers
13) trouble report
14) check valves
15) meter reader
16) meter reading
17) installed backward
18) meter reading
19) pressure from house
20) re-read meter complaint
21) field observations
22) read book
23) we read them
24) meter reading
25) electronic readings
26) work order
27) meter repair shop
28) low reading
29) fire sys/detector meter scv1" stuck open

7. Have you documented that fewer meters have been running backwards since you initiated a cross-connection control program?
(311)
8.04% - Yes
25.40% - No
27.65% - Don't Know
38.91% - The question does not apply

8. Does your system have a regular flushing program?
(350)
80.57% - Yes
19.43% - No

9. If your system has a regular flushing program, how often is the system flushed?
(276)
Every 10.26 months.

10. Have you experienced any water quality complaints, problems or other water system operational problems as a result of flushing activities?
(325)

63% - Yes
37% - No

11. If yes what concerns were experienced?

- 40
- air
- air in water
- air/cloudy water
- borwn water
- brown color water
- brown water
- brown water
- brown water/low pressure
- chlorine taste
- cloudy water
- cloudy water
- cloudy water
- cloudy water
- cloudy water
- cloudy water
- cloudy water
- cloudy water
- cloudy water, black specs, sand, low pressure while flushing
- cloudy water, high chlorine taste
- cloudy water, low pressure
- cloudy, discolored water
- color
- color
- color
- color
- color
- color
- color and sediment
- color due to iron sediment
- color in water, air in water
- color of water
- color, appearance, short supply volume
- color, low pressure
- color, silt in water
- colored water
- colored water
- Colored water
- colored water
- colored water
- colored water from mineral sediment with the pipeline
- colored water, clogged meter
- colored water, rust
- damage to downstream properties
- developing flushing program - rust & turbidity
- dirty
- dirty
- dirty
- dirty
- dirty or cloudy water
- dirty or rusty water
- dirty water
- dirty water
- dirty water
- dirty water
- dirty water
- dirty water
- dirty water
- dirty water
- dirty water
- dirty water
- dirty water
- dirty water
- dirty water
- dirty water
- dirty water
- dirty water
- dirty water
- dirty water
- dirty water
- dirty water
- dirty water
- dirty water
- dirty water
- dirty water calls
- dirty water complaints
- dirty water complaints if we turn on or off too fast
- dirty water concerns before flushing was actually conducted
- dirty water while flushing
- dirty water, air in water
- dirty water, dirty laundry, pressure fluctuation, miminal complaints
- dirty water, low pressure
- dirty water, low pressure
- dirty water, plugged screens
- dirty, sandy water due to flushing activities
- dirty, taste and odor
- dirty/cloudy
- dirty/cloudy water, low pressure, run off to driveways
- discoloration
- discoloration
- discoloration
- discoloration of water
- discolored and or milky water
- discolored water
- discolored water
- discolored water
- discolored water
- discolored water
- discolored water
- discolored water

- discolored water
- discolored water
- high manganese
- increased IDS - sand etc.
- increased manganese after flushing
- iron
- iron and manganese
- iron and manganese in the water
- iron, cloudy, mainbreaks
- it has not always sitrred Fe up in some lines
- low pressure
- low pressure
- low pressure during flushing
- low pressure when flushing
- low pressure, dirty water
- low pressure, dirty water
- low pressure, rusty water
- low pressure, sediment
- low pressure/dirty water
- mostly sand and calcium deposits
- movement of sludge and sand
- muddy water
- muddy water
- pipe scale in areas inadequately flushed
- pressure problems, color, odor
- prior to establishing a regular flushing program we experienced cloudy water complaints
- red water
- red water
- red water
- red water
- red water
- red water
- red water, black water
- red water, stained clothes in washer
- rocks in meters
- rust
- rust
- rust colored water fro a couple of days
- rust in water after flushing
- rusty
- rusty
- rusty water
- rusty water
- rusty water
- rusty water
- rusty water
- rusty water
- rusty water
- rusty water
- rusty water
- rusty water
- rusty water
- rusty water
- rusty water
- rusty water
- rusty water
- rusty water
- rusty water
- rusty water
- rusty water
- rusty water
- rusty water, sediment
- sand
- sand in faucet screens
- sediment
- sediment
- sediment
- sediment, color, taste, odor, pressure, wasting water
- small particles, odor
- some hydrants need repairs
- stirred up particulates
- taste
- temp. dirty water
- temporary loss of pressure
- temporary red wtaer from iron
- toilets had discolored water
- turbid water
- turbidity
- turbidity
- turbidity
- turbidity - color
- wash outs from flow of water from hyds
- wasted water, rusty water
- water aesthetics
- water quality complaints
- we flush due to rusty water
- yellow water

---

12. What percentage of your service connections are (should add up to 100%) (1828)

37.98% - ½" connections (48)
68.15% - ¾" connections (271)
35.30% - 1" connections (280)
3.32% - 1 ½" connections (222)
5.55% - 2" connections (261)
1.3% - 3" connections (155)
1.29% - 4" connections (191)
1.86% - 6" connections (175)
1.35% - 8" connections (106)
1.07% - 10" connections (62)
0.86% - 12" connections (29)
Other connections (28)

- 16"
- 11% - 5/8"
- 95% - 5/8"
- 1%
- 95% 5/8"
- 92.3% 5/8"
- 13% - 1/ 1/4"
- 79 - 5/8
- 94.40% 5/8"
- 80% 5/8"
- 5
- 86% 5/8"
- 11.98% - 5/8"
- 66.3 - 5/8"
-
- 85% 5/8"
- 2
- 95% - 5/8"
- 95% 5/8"
- 25% - 5/8"
- .5
- 94.286% - 5/8"
- 9/158 5/8"
- 87% 5/8"
- 46% 5/8"
- 66 - 5/8"
- 72% - 5/8"
- 67% - 5/8"
- 2

13. In the initial questionnaire question 24 read, "If your agency uses system protection, how many of each backflow preventer are installed in your system for system protection?" The options were.

- Air gap
- Reduced pressure principle assemblies
- Reduced pressure principle detector assemblies
- Double check valve assemblies
- Double check detector assemblies
- Pressure vacuum breaker assemblies
- Atmospheric vacuum breaker assemblies
- Atmospheric vacuum breaker assemblies
- **OTHERS**
- Not applicable.

If you marked (or would mark) "Others," please specify what other types of backflow preventers are used.
(22)

- air gap/double check
- all new household connections have mueller dual checks at the meter
- check valves in certain meters
- county commissioners will not enact backflow program
- double check
- double check on premise. If hazard is higher RP
- dual check
- dual check
- dual check setters
- dual check valves at meters
- dual check with vent
- dual checks
- dual checks
- dural check
- hose bib vacuum breakers
- no requirement at this time. Rules and regs being updated
- resi dual checks
- residential dual checks
- residents have dual checks
- RP or DCDA only
- rpz on all commercial meters, air gaps on drop hoses for filling tanks
- single check valves on hydrant meters

14. In the initial questionnaire question 31 read, "What method of enforcement do you consider to be the most effective?" If you selected "Termination of service," could you please clarify?
(311)

4% - The water service is actually turned off
33% - The customer is notified that the water services will be turned off unless he/she complies with your requirements.
63% - Both of the above

15. Question 10 of the initial questionnaire stated, "How many water connections are there in your system in each of the following categories?" Could you also provide the number of these that currently have backflow protection at the service connection?
(1519)

Total number of water connections:
(1) 2387.88 Commercial/Industrial Domestic
(2) 21253.4 Residential Domestic
(3) 884.68 Irrigation
(4) 733.02 Fire suppression

Number of those with backflow protection

(1)	728.58	Commercial/Industrial Domestic
(2)	1677.6	Residential Domestic
(3)	403.02	Irrigation
(4)	389.67	Fire suppression

16. Has the AwwaRF report entitled, "Impact of Wet Pipe Fire Systems on Drinking Water Quality" had any effect on the way your agency addresses backflow requirements for fire protection water service connections?
(348)

11% - Yes
42% - No (go to Question 18)
47% - Haven't seen the report (go to Question 19)

17. If yes, what effect or changes resulted due to the information contained in the report?

- all new fire sprinkler systems require an approved USC-FCCCHR backflow assembly
- better comm. With fire dist. Procedures and our desire to continue installing DCDA as system protection
- changed contract specs
- currently working on specs on fire system protection
- double check detectors required
- existing class 1 fire sys are not retrofitted
- eye opening info changes needs to be done with state plumb. Code
- fire line always has to have a DCDA or DC
- FIRE SPRINKLER SYSTEM MUST PROVIDE BACKFLOW PROTECTION
- focus on hazard
- helped me convince fitter & eng. That alarm checks fail
- implementation of CC on exist. Systems w/miminal press. Is not done unless they're redesigned
- in the near future we will require BF on class I and II
- increased inspections of services and require DC on new services
- minimum requirements of a double check device
- new requirements in the 1997 tax
- no more single detector checks for firelines. Double detector checks are minimum allowed now.
- now require RP units
- reduced our efforts on systems without anti-freeze
- require existing systems w/o b.f.p. to install a device
- require used of testable single check valves on exist. Fire systems rather than double check valves
- separating fire service from regular service
- substantiate protection on fire systems, report is misleading to others
- The DWWD requires DCDC on all fire services
- upgrades in backflow prev.

- use data to educate consumers
- verification of high hazard protection for glycols & such
- we are currently working with the fire dept to cover new fire services
- we require backflows on all firelines
- we required approved bfp on fire systems before the report was completed
- we still require backflow on dedicated fire services that are part of new construction, however, it will effect how we deal with retrofits
- we use the recommendations to help ed. The users
- we were requiring backflow protection before the AwwaRF study`
- went back to installing d.c. on all new installations
- yes, we don't believe the findings, we don't want this water back in our system, so devices are installed

18. If no, please state the reasons you feel the recommendations of the report are not being followed by your agency.
(134)

48% - Agency policies already consistent with recommendations.
12% - Conflicts with local codes/regulations
16% - Fire protection systems not considered a hazard
25% - Other

- all ready were doing
- backflow protection was required before the report
- city did not accept IAPMO in entirety
- conflict with IPC
- consideration will be made when class attended Oct. '00
- doing fire systems per the CA plumb. Code
- don't have a copy
- each is looked at and most require RPZ for health reasons
- EPA codes and regs
- expense
- Have not addressed this concern
- have not seen the report
- haven't read it
- haven't read it
- haven't received report
- haven't seen it
- haven't seen the report
- heven't read it
- I don't know. AL doesn't have a standard
- no fire protection
- not aware of report
- not responsible for this
- our policies are more restrictive in some cases
- report never viewed
- rural setting only 5 fire sys. All protected and all uncharges sprink. Sys.
- seen report, we go by Pac/NW Section
- state drinking water ccc regulations enforced a minimum
- state law
- there are still conn. With compound meters and no BFP
- we are water wholesalers
- we don't have a problem in this area
- we have none in our system
- we have none of those systems

19. Number 16 in the original questionnaire asked if you conducted site surveys. If you answered "No" to Number 16 in the original questionnaire, what is being done in lieu of site surveys?
(411)

77.86% - Blanket program, we require backflow preventers on all (check all that apply)

(92) Commercial services
(28) Residential services
(81) Fire services
(76) Irrigation services
(43) Other services______________

- air gap
- blanket
- blanket
- blanket
- blanket
- blanket
- blanket
- blanket
- blanket
- blanket
- blanket
- blanket
- blanket
- blanket
- blanket
- blanket
- blanket
- blanket
- blanket
- blanket
- blanket
- blanket
- blanket
- blanket
- blanket
- blanket
- blanket program
- boilers and possible hazards
- booster pump requires b.f.p.a.
- compliance
- construction
- heating & ind. Process systems
- industrial
- industrial
- mfg.
- municipal connections
- plating plants, phoo labs, hospitals, sewage pumps, mortuaries
- residential when rec. is avail.
- temp & water haulers
- under ordnance
- water dist
- wells
- WWTP

5.6% - Site surveys are conducted by another administrative authority
3.41% - Site surveys are conducted by a third party for our agency
2.43% - Survey conducted by mail in questionnaire
0.97% - Survey conducted by phone
4.14% - Other________________________

- backflow testers do not notify us of problems
- building plan review
- code enforcement & public works
- district (ccs)
- no time to do this
- plan examination
- plumbers
- property has alternate water source
- site surveys are planned when resources exist
- site surveys have been done on those necessary
- specific classified accounts
- specific high risk areas
- survey by bldg. Permits obtained
- survey in disclosure packet @ time home puch.
- we do site surveys
- we do site surveys
- X-conn ctrl insp. Does site inspections

5.6% - Nothing else is being done.

20. Question 35 in the original questionnaire asked, "Does your cross-connection control program coordinate efforts with any of the following?

- State Health Department
- Local Health Department
- Other local Water Agencies
- Plumbing Code enforcement agency
- Fire Department
- Other
- Organizations (AWWA, ABPA, etc.)

Specifically what type of coordinated effort exists?
(412)

16.26% - Joint site surveys
49.27% - Regulatory support
27.43% - Enforcement actions
7.04% - Other_______________

- Certification of Testers
- city atty's office, dev. Dept.
- communication, review of issues, etc.
- coordinate test months for customers w/meter and internal devices
- coordinatioin of training
- CSA AWWA
- education, notification
- information and education
- information exchange
- Inspec. Does the inspec., util handles paperwork
- joint water system meetings
- local & state DOH review of plans
- none
- none
-
- none
- plan examination
- public education materials
- questionnaire to customer
- regulatory guidance
- reports - meetings
- reuse dept.
- state dept of env. Service
- twrcc
- utility and pumbing code enforcement
- VDH reviews our annual survey
- we must give them an annual report
- we provide info to the agencies above
- when asked by d.e.q.
- work under same manager

21. Question 40 of the original questionnaire stated, "How many incidents has your agency experienced since 1970?" To get a better understanding could you please complete the following, by filling in the number of incidents that have occurred for each period of time indicated?
(130)

4.3 - from 1971 through 1975.
3.10 - from 1976 through 1980.
2.2 - from 1981 through 1985.
4.43 - from 1986 through 1990.
3.13 - from 1991 through 1995.
2.66 - from 1996 through 2000.

22. In what year was your cross-connection program implemented (Please provide the year only).
(488)

(250) on paper (adopted legal document)

1941	1960	1970	1972	1973	1975	1976
1944	1961	1970	1972	1974	1975	1976
1951	1967	1971	1973	1974	1975	1977
1953	1969	1972	1973	1975	1975	1977

1977	1982	1986	1989	1991	1993	1997
1977	1982	1986	1989	1991	1994	1997
1977	1982	1987	1989	1991	1994	1997
1978	1982	1987	1989	1991	1994	1997
1978	1983	1987	1989	1991	1994	1997
1978	1983	1987	1989	1992	1994	1997
1978	1983	1987	1989	1992	1994	1998
1978	1983	1987	1989	1992	1994	1998
1978	1983	1987	1989	1992	1994	1998
1978	1983	1987	1989	1992	1994	1998
1978	1984	1987	1989	1992	1994	1998
1978	1984	1988	1989	1992	1994	1998
1979	1984	1988	1989	1992	1995	1998
1979	1984	1988	1989	1992	1995	1998
1979	1984	1988	1989	1992	1995	1998
1979	1985	1988	1989	1992	1995	1998
1980	1985	1988	1989	1992	1995	1999
1980	1985	1988	1990	1992	1995	1999
1980	1985	1988	1990	1992	1996	1999
1980	1985	1988	1990	1993	1996	1999
1980	1985	1988	1990	1993	1996	1999
1980	1985	1988	1990	1993	1996	1999
1980	1985	1988	1990	1993	1996	1999
1980	1985	1988	1990	1993	1996	1999
1980	1985	1988	1991	1993	1996	1999
1980	1985	1988	1991	1993	1996	1999
1981	1986	1988	1991	1993	1996	2000
1981	1986	1988	1991	1993	1996	2000
1981	1986	1988	1991	1993	1997	2000
1981	1986	1988	1991	1993	1997	2000
1981	1986	1988	1991	1993	1997	
1982	1986	1988	1991	1993	1997	

(238) in practice.

1938	1975	1980	1983	1985	1988	1989
1960	1975	1980	1984	1985	1988	1989
1960	1976	1980	1984	1986	1988	1990
1961	1977	1980	1984	1986	1988	1990
1964	1977	1980	1984	1986	1988	1990
1968	1978	1980	1984	1986	1988	1990
1970	1978	1980	1984	1986	1988	1990
1970	1978	1980	1985	1986	1989	1990
1970	1978	1981	1985	1987	1989	1990
1971	1978	1981	1985	1987	1989	1990
1972	1978	1981	1985	1987	1989	1990
1972	1979	1981	1985	1987	1989	1990
1974	1979	1981	1985	1987	1989	1990
1975	1979	1982	1985	1987	1989	1990
1975	1979	1982	1985	1987	1989	1991
1975	1980	1982	1985	1988	1989	1991
1975	1980	1982	1985	1988	1989	1991
1975	1980	1982	1985	1988	1989	1991
1975	1980	1983	1985	1988	1989	1991

1991	1993	1994	1995	1996	1997	1999
1991	1993	1994	1995	1996	1997	1999
1992	1993	1994	1995	1997	1998	1999
1992	1993	1994	1995	1997	1998	1999
1992	1993	1994	1996	1997	1998	1999
1992	1993	1994	1996	1997	1998	1999
1992	1993	1994	1996	1997	1998	1999
1992	1993	1994	1996	1997	1998	1999
1992	1993	1995	1996	1997	1998	1999
1992	1993	1995	1996	1997	1998	1999
1992	1994	1995	1996	1997	1998	2000
1993	1994	1995	1996	1997	1998	2000
1993	1994	1995	1996	1997	1998	2000
1993	1994	1995	1996	1997	1998	2000
1993	1994	1995	1996	1997	1998	2001

23. What approval or listing requirement is there for backflow prevention assemblies in your area of jurisdiction? (If there is a combination of these, please indicate all that apply to your area.)
(410)

45.85% - University of Southern California
11.95% - American Society of Sanitary Engineering
7.32% - International Association of Plumbing and Mechanical Officials
2.68% - Canadian Standards Association
9.27% - Underwriters Laboratory
4.15% - Factory Mutual
18.78% - Other______________________________

ABPA
ABPA, TN Back. Prev. assoc.
AWWA
awwa
AWWA
AWWA
AWWA
AWWA
AWWA
AWWA approval
AWWA standards
CA dept of health services
CA dept. of health serv.

CA Health Dept.
CA State DHS
Cal. DHS
cdhs
CO CCC manual
DEHEC
dept of natural resources
Dept. of Health Serv.
Dept. of Health Serv.
dept. of healther
dhs
dhs
district list
IEPA

LA Cnty.
LA Cnty. Health Dept.
local ordnance
MA DEP
Madera county env. Health?
MDNR
MI DEQ
MI xxx rules manual
MN dept of health
MO Dept. of Natural Resources
MO state DNR

Nassau county health dept
NV state health div.
NY state approval
NY state dept. of health
NY State dept. of health
NY State DOH
NYS sanitary code
Ohio Env. Pro. Agency
Ohio EPA
Ohio EPA

OR Health Dept. AWWA
OR Health Div. OR state health
Oregon Health Division
our const. Specs
plumb. Code & IEPA
Riverside health dept.
SCDHEC
sdhs
So. Fla. Bldg. Code M-14

state
state
state dept.
state DOH
state health
state health
state health dept
state law
state of CA
state of calif.

State of Utah	state regs	VDH regs
state of utah DEQ	twrcc	WA State approved list
state plumb. Inspector	ULC	Washington D.C.

24. How does your agency provide water? (Mark all that apply.)
(436)

29.82% - as a wholesaler
62.16% - as a retailer
4.82% - to agricultural systems
3.21% - Other.

25. What is the average cost in your area of jurisdiction for each of the following for each size of ***reduced pressure principle assembly*** indicated:
(901)

	1"	4"	10"
Testing an assembly?	$56.78	$67.14	$69.41
Repair of assembly?	$ 90.31	$207.55	$ 552.57
Installation?	$ 343.67	$1935.04	$5204.13
Purchase?	$159.57	$1241.92	$4839.06

26. For field testing backflow preventers which does your agency use?
(369)

35.77% - In-house testers (i.e., employed by administrative authority)
64.23% - Private testers (i.e., commercial testers)

27. Does your agency operate a recycled water system?
(336)

10.71% - Yes
89.29% - No (go to Question 32)

28. If yes, in Question 27,do the recycled water services require back flow protection?
(40)

52.50% - Yes
47.50% - No (go to Question 32)

29. If yes, in question 28, what type of backflow protection is required?
(25)

12% - Double check valve assembly
72% - Reduced pressure principle assembly
4% - Pressure Vacuum Breaker Assembly
12 - Other____

air gap
air gaps
air gap

30. If your agency operates a recycled system, do the potable water services to sites having recycled water require backflow protection?
(42)

73.81% - Yes
26.19% - No (go to Question 32)

31. If yes, in Question 30, what type of protection?
(33)

9.09% - Double check valve assembly
78.79% - Reduced pressure principle assembly
0% - Pressure Vacuum Breaker Assembly
12.12% - Other____

air gap
air gap
air gap
no direct - swinging L

32. If you indicated in the first questionnaire that you have documented backflow incidents would you be willing to provide copies of the documentation?
(265)

3.02% - Yes (documentation is included with this questionnaire)
6.79% - Yes (documentation will be sent under separate cover)
14.34% - No, documentation will not be provided.
75.85% - There is no documentation of backflow incidents

33. How many water quality complaints has your agency received over the past ten years?

52.47 – 2000 (137)
84.35 – 1999 (158)
89.03 – 1998 (144)
86.47 – 1997 (130)
98.17 – 1996 (113)
84.39 – 1995 (98)
109.43 – 1994 (87)
77.65 – 1993 (82)
80.40 – 1992 (75)
59.50 – 1991 (78)

34. If you have a cross-connection control program, do you feel the program has had an effect on water quality complaints and problems?
(324)

29.32% - Yes
33.02% - No (go to Question 36)
37.65% - Not sure

35. If yes, please describe the effect your program has had on these complaints.
(84)

1) all high hazard and most moderated hazard facilities are contained

2) assurance that known chemical locations are isolated
3) Bacflow prev. assys. Prevented contamination to our system when our system went to 0 PSI after a Western U.S. power outage
4) because now the ccc dept is notified of w.q. complaints, inves. Now done by ccc dept.
5) better assurance of a clean water supply
6) better public health, better system integrity
7) community and employee awareness
8) complaints held to a minimum
9) customers are more aware of x-conn. & will retrofit more readily because of known risks
10) customers feel more confident of water qual. When they are educated on the effects of a backflow preventer
11) decreasing numbers of reported incidents
12) do I need a bacflow device? Why? Why do I have to pay? What about my neighbor?
13) education awareness
14) education of homeowners
15) education/elimination
16) eliminated pop dispensor related taste complaints. Reduced options to be investigated.
17) elimination of sources for contamination
18) few if any complaints and no reported backflow incidents
19) fewer complaints
20) fewer complaints as more people understand purpose of program and comply with us
21) gives better security to customer
22) gives customer peace of mind
23) has eliminated actual or potential cross connections
24) Has stopped reverse flows on newer services that may have caused system problems.
25) have inspection done in flushing area
26) have no records of backflow problems
27) have zero backflow incidents
28) hopefully eliminated life threatening complaints
29) hopefully prevented cross-connection
30) if we find a violation we fix it before someone gets sick
31) in a city of this size without any active ccc program, there are undoubtedly numerous conditions that would cause backflow to occur.
32) inspect bldg. With permits, joint surveys with county health. No complaints
33) inspections usually find internal pumbing problems that cause complaints
34) It has protected our system from major contaminates
35) keeping the customers water from re-entering the potable system
36) lab personnel can use ccc records to assist in determining causes
37) less complaints, more easily determined, course of action
38) less cross connections
39) less or minimal w.q. complaints
40) minimize backflow throught containment, education, enforcement
41) minimizes draining of private services during high flows assoc. with flushing / fire fighting
42) no back siphonage into system
43) no complaints due to cross connections
44) no direct evidence to support feeling
45) normal household maint. - hot water tank, cln strainers, routine plumb.
46) on site inspections
47) our program cannot effect muddy water complaints

48) people fee safer & secure when we talk about it
49) people realize that we are making an effort to protect them and are more tolerant
50) people realize we are doing more to provide good water quality
51) people see their water company is protecting their water
52) pressure loss in some isolated areas
53) preventative
54) preventative maintenance
55) prevented possible incidents
56) program has reduced potential for cust. Contaminants to enter our system. Potential for complaints therefore reduced.
57) program reassures both consumers and w.q. staff that complaint is not x-xonn incident
58) program too new
59) provided assurance to people & kept stale water from backflowing
60) PUBLIC EDUCATION, MORE INFORMATION
61) reduced complaints
62) reduced complaints and problems
63) reduced many of the complaints
64) reduced them
65) reduction on water qual. Complaints
66) referred to cc program
67) residents feel all is being done to deliver quality water
68) stop thermal expansion
69) sys. Is adequately protected and customers are better educated
70) taste and odors
71) the program allows us to verity that a water qual. Complaint is not due to a backflow incident
72) very few
73) very little, most complaints were home siphoning related
74) we are better prepared to handle complaints, better preparded to respond to emergencies
75) we assume that we have prevented backflow incidents
76) we believe the number of wtr. Qual. Complaints would be higher if we didn't have a ccc program
77) we have had no bad coliform samples since we started the program
78) we have received less w.q. complaints
79) we have seen in an anectodal sense is a decrease in the occurrence of backflow events
80) we haven't had any incidents
81) with the program the public is aware that backflow problems can effect water quality issues.
82) wtr qual has improved, complaints down, sand and pressure main problems
83) zero cross connection problems

36. What percentage of your agency's annual Operations & Maintenance budget is directed towards cross-connection control?
(230)
3.65%

37. Does your agency have a Membership in or financially support any of the following organizations (check all that apply)?
(699)

48.5% - AWWA (American Water Works Association)

14.31% - USC (University of Southern California Foundation for Cross-Connection Control and Hydraulic Research)

14.31% -AwwaRF (Awwa Research Foundation)

14.74% - ABPA (American Backflow Prevention Association)

8.15& - Other____

• apwa swana	• NRWA, FRWA
• ASSE	• NY rural water assoc
• asse, aspe	• NYRW
• asua, awpca	• OCCIRS
• Awwa	• OH rufal water assoc.
• Bcwws	• Oregon Assoc. Water Util.
• BEWA	• PRWA
• Bluegrass cc prev assoc	• regional water groups
• CA rural water assoc.	• rural water
• CWEA	• rural water assoc.
• I.A.M.U.	• small water users assoc.
• Ill rural water assoc.	• SRC4
• IRWA	• src4/sajb wellhead protection
• LA county x-conn inspectors assoc.	• state chapter abpa
• local abpa	• TN Backflow Prev. Assoc.
• local bfpa	• TN Sec. Of backflow
• Long island water conf.	• TX water util. Assoc.
• Maine BPA	• VA ccc assoc
• Mich rural water	• VA CCC Assoc.
• MRWA	• VA ccc assoc.
• Nantional rural water	• VA rural water assoc.
• National rural water	• virginia cross connection control assoc
• NCBPA	• Water works operators assoc.
• Ncbpa	• wcwwa
• New England WWA, NE Rural wtr assoc.	• WI rural water assoc.
• NEWWA, MWWA, PCWWA	• WRWA
• NEWWA, NERWA	• PRWA
• Nor Cal back prev assoc	• regional water groups
• North Carolina rural water assoc.	• PRWA
• NRWA	•

APPENDIX F
SYSTEM CHARACTERISTICS FOR MODEL CITY B

94	V	J675	102.9	198.9	96.0	-68.5	55.9
95	J	J649	102.9	202.1	99.2	196.6	0.0
96	V	J647	93.8	211.1	117.3	154.6	24.8
97	V	J646	87.0	211.1	124.1	6.4	17.3
98	J	J642	87.0	211.6	124.6	-108.8	0.0
99	V	J633	98.0	212.4	114.4	52.2	43.5
100	V	J612	82.7	216.5	133.8	-282.0	82.5
101	V	J611	73.2	218.8	145.6	-427.5	38.3
102	V	J609	72.5	220.3	147.8	-391.6	42.0
103	V	J587	88.7	223.9	135.2	-243.5	105.0
104	V	J526	111.1	229.9	118.8	83.5	94.2
105	J	J67	73.4	290.1	216.7	738.0	0.0
106	V	J195	73.0	279.0	206.0	431.9	21.0
107	V	J199	73.8	269.0	195.2	321.5	18.0
108	V	J318	79.9	266.7	186.8	243.7	6.0
109	V	J319	76.9	266.5	189.6	90.2	5.3
110	V	J485	96.5	205.8	109.3	599.0	21.0
111	V	J234	95.1	240.6	145.5	194.8	41.3
112	V	J153	79.0	283.7	204.7	112.4	21.8
113	V	J149	90.0	276.2	186.2	322.9	13.5
114	V	J148	103.5	269.8	166.3	230.7	28.5
115	V	J269	115.4	227.3	111.9	0.0	49.5
116	V	J268	111.9	227.6	115.7	-66.3	18.8
117	V	J274	105.0	229.5	124.5	18.8	18.8
118	V	J297	87.0	235.0	148.0	-178.7	20.3
119	V	J302	84.2	244.2	160.0	-219.0	62.5
120	J	J305	120.5	249.7	129.2	70.9	0.0
121	V	J309	127.0	250.7	123.7	142.8	494.8
122	J	J314	120.0	254.0	134.0	0.0	0.0
123	V	J315	87.5	260.1	172.6	20.1	40.4
124	V	J316	83.0	263.0	180.0	-158.5	6.0
125	V	J317	86.3	263.7	177.4	72.2	9.8
126	V	J321	77.7	263.2	185.5	147.3	9.0
127	V	J346	76.5	258.8	182.3	138.3	26.3
128	V	J417	82.0	235.2	153.2	65.0	47.3
129	V	J514	82.5	231.2	148.7	240.9	39.0
130	V	J513	78.9	230.9	152.0	242.2	24.0
131	V	J519	81.8	227.7	145.9	218.2	24.0
132	V	J564	80.0	227.2	147.2	43.4	43.5
133	V	J563	75.5	227.2	151.7	-55.5	24.0
134	V	J518	89.5	227.5	138.0	90.0	90.0
135	V	J537	112.0	239.3	127.3	92.5	29.3
136	V	J544	95.2	239.0	143.8	63.2	27.0
137	V	J451	78.4	238.8	160.4	36.2	49.5
138	V	J607	111.0	223.3	112.3	29.3	29.3
139	V	J624	92.0	212.4	120.4	81.5	23.3
140	V	J748	98.7	211.4	112.7	14.3	14.3
141	V	J762	87.7	209.6	121.9	15.4	15.8
142	J	J763	91.5	209.6	118.1	1.1	0.0
143	V	J287	104.9	221.7	116.8	-34.8	33.0
144	V	J491	86.6	222.9	136.3	14.3	14.3
145	V	J479	118.0	199.4	81.4	70.9	28.5
146	V	J652	100.3	201.3	101.0	348.7	18.8
147	V	J651	100.7	200.5	99.8	286.5	286.5
148	V	J653	112.3	198.4	86.1	172.0	12.0
149	V	J654	116.0	196.8	80.8	9.4	5.3
150	J	J657	94.4	194.4	100.0	26.3	0.0

151	V	J695	90.5	194.1	103.6	49.7	18.0
152	V	J696	89.9	194.1	104.2	43.1	12.0
153	V	J698	90.4	194.0	103.6	30.7	22.5
154	V	J701	98.1	194.0	95.9	23.8	7.5
155	V	J678	108.2	198.6	90.4	111.8	27.0
156	J	J679	105.0	198.5	93.5	84.8	0.0
157	V	J683	102.7	198.4	95.7	49.1	9.0
158	V	J684	101.0	198.4	97.4	92.7	15.0
159	V	J673	96.8	211.0	114.2	123.5	19.5
160	J	J672	100.5	211.0	110.5	0.0	0.0
161	J	J674	98.7	210.9	112.2	131.3	0.0
162	V	J745	86.8	210.6	123.8	131.3	9.0
163	V	J744	85.8	210.6	124.8	114.9	14.3
164	V	J741	87.8	210.0	122.2	13.5	13.5
165	V	J731	110.2	197.6	87.4	95.7	18.8
166	V	J727	122.7	194.4	71.7	123.4	41.3
167	V	J705	117.0	194.3	77.3	46.9	16.5
168	J	J704	102.0	194.0	92.0	53.5	0.0
169	V	J702	96.0	194.0	98.0	46.7	17.3
170	J	J706	92.5	193.9	101.4	29.4	0.0
171	V	J707	91.2	193.9	102.7	29.4	24.8
172	J	J711	91.6	193.9	102.3	18.7	0.0
173	V	J713	90.7	193.9	103.2	18.7	17.3
174	V	J717	119.9	193.9	74.0	36.6	22.5
175	V	J726	113.0	193.9	80.9	155.5	55.5
176	V	J94 EOL	110.0	193.0	83.0	100.0	100.0
177	V	J194	79.3	275.1	195.8	100.2	10.5
178	V	J202	81.5	265.4	183.9	27.8	27.8
179	V	J145	93.0	272.1	179.1	78.7	11.3
180	V	J97	96.0	272.1	176.1	0.0	0.0
181	V	J146	99.0	269.8	170.8	67.4	17.3
182	V	J147	100.0	269.0	169.0	102.6	32.3
183	V	J141	112.9	263.6	150.7	99.8	99.8
184	V	J96	112.6	263.6	151.0	0.0	0.0
185	V	J171	115.0	259.3	144.3	120.2	75.0
186	V	J169	105.0	259.3	154.3	147.9	117.8
187	V	J163	103.3	258.7	155.4	30.1	39.0
188	V	J191	100.0	258.7	158.7	112.1	12.8
189	V	J189	105.0	256.4	151.4	146.3	32.3
190	V	J187	111.0	252.2	141.2	284.1	339.0
191	J	J186	119.0	253.0	134.0	35.6	0.0
192	J	J216	134.8	252.0	117.2	35.6	0.0
193	J	J222	127.5	251.6	124.1	35.6	0.0
194	J	J214	122.0	251.2	129.2	0.0	0.0
195	V	J134	126.5	262.9	136.4	27.0	27.0
196	V	J206	97.0	264.7	167.7	38.3	38.3
197	V	J322	94.4	248.8	154.4	138.1	95.4
198	V	J323	96.0	248.5	152.5	25.1	8.3
199	V	J324	115.5	249.4	133.9	89.4	17.3
200	V	J342	103.0	245.2	142.2	61.2	83.1
201	V	J344	96.4	248.4	152.0	64.7	0.0
202	V	J367	90.7	243.8	153.1	64.7	44.3
203	V	J368	96.9	244.1	147.2	27.2	78.8
204	V	J374	107.0	244.8	137.8	45.4	27.8
205	V	J378	108.0	244.7	136.7	66.8	66.8
206	V	J434	101.7	242.5	140.8	101.7	42.8
207	V	J439	83.4	239.5	156.1	72.5	72.5

208.	V	J454	102.0	239.0	137.0	27.8	27.8
209	J	J605	122.0	232.5	110.5	104.2	0.0
210	V	J304	100.0	248.9	148.9	41.7	41.7
211	V	J295	87.0	234.9	147.9	19.9	21.8
212	V	J405	99.9	234.9	135.0	15.8	15.8
213	V	J523	83.8	230.0	146.2	23.3	23.3
214	V	J756	112.0	196.5	84.5	340.9	340.9
215	V	J623	97.9	213.0	115.1	12.8	12.8
216	J	J632	83.1	212.4	129.3	0.0	0.0
217	J	J761	95.0	209.6	114.6	0.0	0.0
218	V	J101 EOL	105.0	245.9	140.9	0.0	7.5
219	V	J102	110.1	246.0	135.9	-7.5	3.0
220	V	J104	111.9	246.0	134.1	18.6	7.5
221	V	J119	118.0	245.8	127.8	37.0	34.5
222	V	J121	113.5	245.8	132.3	23.1	45.0
223	V	J123	106.0	247.0	141.0	11.3	11.3
224	V	J242	113.5	235.8	122.3	105.8	24.8
225	V	J236	110.0	236.0	126.0	134.7	36.0
226	V	J262	113.9	230.1	116.2	99.9	57.0
227	V	J263	110.0	230.2	120.2	74.3	7.5
228	V	J264	109.0	229.0	120.0	13.8	12.0
229	V	J265	113.7	229.0	115.3	4.5	4.5
230	J	J292	114.0	222.2	108.2	-18.4	0.0
231	J	J289	111.9	222.3	110.4	0.0	0.0
232	V	J95 EOL	91.5	200.6	109.1	10.4	10.4
233	V	J658	96.5	194.5	98.0	-23.4	6.0
234	V	J662	110.9	196.8	85.9	-39.7	16.5
235	V	J663	110.9	198.1	87.2	52.0	9.0
236	V	J664	111.9	197.8	85.9	11.4	12.0
237	J	J665	113.4	197.8	84.4	-0.6	0.0
238	V	J666	96.9	198.5	101.6	50.0	19.5
239	V	J667	97.8	198.4	100.6	13.5	13.5
240	V	J685	101.8	198.0	96.2	77.7	15.8
241	V	J686	114.0	197.6	83.6	46.7	6.8
242	J	J690	123.7	196.9	73.2	0.0	0.0
243	V	J692	123.4	197.3	73.9	12.0	12.0
244	V	J694	93.8	194.2	100.4	37.1	9.8
245	V	J697	95.2	194.1	98.9	12.0	12.0
246	V	J736	112.0	198.0	86.0	24.8	24.8
247	V	J708	104.5	193.9	89.4	9.0	9.0
248	V	J619 EOL	110.3	214.0	103.7	0.0	0.0
249	V	J61 F ST	-5.5	291.3	296.8	-11436.5	0.0
250	P	J31	17.0	292.0	275.0	-11436.5	-11436.5
251	V	J93	107.0	191.2	84.2	440.6	440.6
252	V	J901	97.0	232.3	135.3	104.2	104.2

****

PIPELINE STEADY STATE DATA

****

LINE	DIAMETER (IN)	LENGTH (FT)	SECTIONS	WAVE SP. (FT/S)	(DARCY)	DISCHARGE (GPM)	HEAD LOSS (FT)
1	24.00	87.	2	3832.	0.02020	4441.33	0.14
2	24.00	1380.	32	3799.	0.03120	9156.83	14.10
3	24.00	700.	16	3854.	0.03130	9304.70	7.41

4	24.00	1500.	34	3886.	0.03440	4607.80	4.28
5	24.00	1700.	39	3840.	0.03440	4598.25	4.83
6	24.00	900.	21	3775.	0.03510	3997.02	1.97
7	24.00	250.	6	3670.	0.03570	3620.51	0.46
8	20.00	700.	16	3854.	0.03720	2575.51	1.68
9	20.00	3450.	78	3896.	0.03720	2575.51	8.27
10	20.00	2075.	47	3889.	0.03710	2751.96	5.67
11	20.00	325.	7	4090.	0.03780	2751.96	0.90
12	20.00	800.	18	3915.	0.03930	1655.33	0.84
13	16.00	1675.	37	3988.	0.03940	1321.30	3.42
14	16.00	1325.	29	4025.	0.02350	1175.53	1.28
15	16.00	575.	13	3896.	0.04080	1022.26	0.73
16	16.00	2500.	55	4004.	0.04240	781.34	1.92
17	14.00	1941.	42	4071.	0.02560	69.80	0.01
18	14.00	1268.	27	4137.	0.02470	-596.65	0.64
19	14.00	325.	7	4090.	0.04380	3565.50	10.46
20	20.00	2793.	63	3905.	0.04940	2364.64	7.50
21	20.00	2605.	59	3889.	0.02300	5087.51	15.07
22	20.00	775.	17	4016.	0.02500	3697.32	2.57
23	20.00	1142.	26	3870.	0.03750	-3672.85	5.61
24	20.00	991.	22	3967.	0.03740	-2919.83	3.07
25	20.00	2678.	60	3931.	0.03710	-2882.11	8.02
26	20.00	3015.	68	3906.	0.03630	-2972.89	9.40
27	20.00	1855.	42	3891.	0.03630	-3167.66	6.57
28	20.00	2100.	47	3936.	0.03570	-3667.51	9.80
29	24.00	1880.	43	3851.	0.03660	-3635.72	3.55
30	24.00	3305.	76	3831.	0.02070	-4445.43	5.28
31	24.00	1315.	30	3861.	0.02060	-4622.67	2.26
32	16.00	1225.	27	3997.	0.11510	1541.61	9.94
33	16.00	2925.	64	4026.	0.11890	1251.81	16.16
34	16.00	280.	6	4111.	0.04360	1049.74	0.40
35	16.00	1330.	29	4040.	0.04490	921.88	1.50
36	14.00	750.	16	4129.	0.04510	830.38	1.35
37	14.00	1450.	31	4120.	0.05640	165.45	0.13
38	14.00	725.	16	3992.	0.04070	872.67	1.30
39	14.00	525.	11	4204.	0.04070	867.72	0.93
40	14.00	2075.	45	4062.	0.04100	810.28	3.23
41	14.00	485.	10	4272.	0.04340	364.63	0.16
42	14.00	1190.	26	4032.	0.04020	1027.12	2.92
43	14.00	2830.	61	4087.	0.04110	886.28	5.28
44	20.00	875.	20	3854.	0.04650	561.38	0.12
45	20.00	425.	10	3744.	0.04750	363.93	0.03
46	20.00	575.	13	3896.	0.04980	255.93	0.02
47	14.00	890.	19	4126.	0.04900	-683.50	1.18
48	14.00	275.	6	4038.	0.04900	-758.50	0.45
49	14.00	430.	9	4209.	0.05620	123.36	0.02
50	14.00	2225.	48	4083.	0.06010	91.06	0.06
51	14.00	820.	18	4013.	0.06010	0.00	0.00
52	20.00	970.	22	3884.	0.02600	1390.19	0.47
53	20.00	1535.	35	3863.	0.02570	1620.97	1.01
54	16.00	1600.	35	4027.	0.02640	198.04	0.05
55	16.00	5100.	112	4011.	0.02970	-96.92	0.04
56	14.00	2175.	47	4077.	0.03880	1183.76	6.84
57	14.00	400.	9	3915.	0.03800	1509.05	2.00
58	14.00	1375.	30	4038.	0.04000	823.02	2.15
59	16.00	725.	16	3992.	0.04870	562.26	0.33
60	16.00	1075.	24	3946.	0.05060	191.68	0.06

61.	16.00	409.	9	3998.	0.05890	59.06	0.00
62	12.00	2832.	60	4158.	0.03200	168.15	0.32
63	12.00	1450.	31	4120.	0.05440	108.89	0.12
64	12.00	2400.	51	4146.	0.06600	108.89	0.23
65	12.00	575.	12	4221.	0.03870	66.30	0.01
66	12.00	2425.	52	4108.	0.04040	40.41	0.02
67	12.00	1970.	42	4132.	0.03870	0.00	0.00
68	12.00	2050.	44	4104.	0.02900	970.07	6.99
69	12.00	525.	11	4204.	0.03470	50.22	0.01
70	12.00	3875.	83	4113.	0.03330	748.56	9.04
71	12.00	1400.	30	4111.	0.06460	-154.04	0.27
72	12.00	495.	11	3964.	0.04810	841.34	2.11
73	12.00	3158.	67	4152.	0.04810	841.34	13.44
74	12.00	2175.	46	4165.	0.04480	505.87	3.12
75	12.00	1825.	39	4122.	0.04590	453.37	2.15
76	12.00	1105.	24	4056.	0.04760	388.07	0.99
77	12.00	575.	12	4221.	0.04790	333.52	0.38
78	12.00	1200.	26	4066.	0.06470	7.91	0.00
79	12.00	480.	10	4228.	0.05930	151.31	0.08
80	12.00	850.	18	4160.	0.06870	-153.65	0.17
81	20.00	200.	5	3524.	0.02470	-4546.05	0.99
82	16.00	584.	13	3957.	0.03060	144.96	0.01
83	10.00	6800.	143	4189.	0.06460	350.34	16.77
84	10.00	1553.	33	4146.	0.03080	260.23	1.01
85	10.00	1495.	31	4248.	0.02960	713.08	7.00
86	10.00	2948.	62	4189.	0.02960	-656.65	11.70
87	10.00	2725.	57	4211.	0.13030	-279.81	8.64
88	10.00	4425.	93	4191.	0.10240	358.93	18.15
89	10.00	700.	15	4111.	0.13060	161.94	0.75
90	10.00	700.	15	4111.	0.08020	60.10	0.06
91	10.00	700.	15	4111.	0.15970	224.63	1.75
92	10.00	1175.	25	4140.	0.16670	174.16	1.85
93	10.00	1500.	31	4263.	0.16970	-33.46	0.09
94	10.00	850.	18	4160.	0.14540	-48.84	0.09
95	10.00	200.	4	4405.	0.17790	124.44	0.17
96	10.00	550.	12	4038.	0.17410	104.32	0.32
97	10.00	2600.	55	4164.	0.12560	290.69	8.58
98	10.00	737.	15	4328.	0.12960	262.29	2.04
99	10.00	1684.	35	4238.	0.14540	5.69	0.00
100	10.00	900.	19	4173.	0.13100	-68.51	0.17
101	10.00	1567.	33	4183.	0.11800	-236.17	3.21
102	10.00	1435.	30	4214.	0.11130	-432.73	9.30
103	10.00	275.	6	4038.	0.13430	154.61	0.27
104	10.00	450.	9	4405.	0.14830	6.36	0.00
105	10.00	875.	18	4282.	0.15620	-108.79	0.50
106	10.00	850.	18	4160.	0.26650	-108.79	0.83
107	10.00	1425.	30	4184.	0.14030	-160.99	1.61
108	10.00	2225.	47	4170.	0.05860	-281.99	3.22
109	10.00	745.	16	4102.	0.05480	-427.49	2.32
110	10.00	600.	13	4066.	0.05500	-391.59	1.57
111	10.00	1450.	30	4258.	0.13100	-243.51	3.50
112	10.00	1825.	38	4231.	0.12330	-293.10	6.01
113	10.00	1875.	39	4235.	0.11690	-376.64	9.67
114	10.00	1500.	31	4263.	0.12660	431.92	11.02
115	10.00	2375.	50	4184.	0.13190	321.54	10.07
116	10.00	900.	19	4173.	0.13740	243.73	2.28
117	10.00	575.	12	4221.	0.15880	90.21	0.23

118	10.00	1050.	22	4204.	0.06590	599.04	7.72
119	10.00	1200.	25	4228.	0.02890	194.78	0.41
120	8.00	614.	13	4161.	0.11710	306.11	6.39
121	8.00	725.	15	4258.	0.12540	193.76	3.24
122	8.00	1520.	31	4319.	0.14570	147.87	4.60
123	8.00	375.	8	4129.	0.11510	322.89	4.27
124	8.00	1040.	21	4363.	0.12110	230.69	6.36
125	8.00	500.	10	4405.	0.16390	-66.27	0.34
126	8.00	925.	19	4289.	0.15040	-118.21	1.84
127	8.00	775.	16	4267.	0.14620	-137.01	2.02
128	8.00	700.	14	4405.	0.15570	-178.75	3.30
129	8.00	1425.	29	4329.	0.14090	-218.96	9.13
130	8.00	1500.	31	4263.	0.14410	-164.43	5.55
131	8.00	800.	17	4146.	0.13560	-235.31	5.70
132	8.00	1269.	26	4300.	0.19250	142.84	4.73
133	8.00	250.	5	4405.	0.13970	-316.32	3.32
134	8.00	400.	8	4405.	0.13970	-316.32	5.31
135	8.00	1700.	35	4279.	0.18500	-51.38	0.79
136	8.00	850.	18	4160.	0.14350	-158.46	2.91
137	8.00	800.	17	4146.	0.16270	-75.34	0.70
138	8.00	850.	18	4160.	0.16080	62.38	0.50
139	8.00	1675.	35	4216.	0.14440	138.29	4.39
140	8.00	1000.	21	4195.	0.14410	-147.53	2.98
141	8.00	3750.	77	4290.	0.05480	42.60	0.35
142	8.00	1120.	23	4290.	0.04660	240.92	2.87
143	8.00	700.	14	4405.	0.05330	102.64	0.37
144	8.00	1475.	30	4331.	0.04760	218.20	3.17
145	8.00	675.	14	4247.	0.05340	117.60	0.47
146	8.00	500.	10	4405.	0.18620	-0.10	0.00
147	8.00	525.	11	4204.	0.05470	76.60	0.16
148	8.00	1600.	33	4271.	0.04950	92.54	0.64
149	8.00	1650.	34	4275.	0.05240	63.24	0.33
150	8.00	3250.	67	4273.	0.05720	36.24	0.23
151	8.00	1500.	31	4263.	0.05770	-190.07	2.97
152	8.00	1350.	28	4247.	0.03100	155.91	0.97
153	8.00	1600.	33	4271.	0.03090	185.28	1.61
154	8.00	800.	17	4146.	0.03910	-43.67	0.06
155	8.00	875.	18	4282.	0.07950	-1.75	0.00
156	8.00	2544.	52	4310.	0.03090	-127.88	1.22
157	8.00	6923.	143	4265.	0.02800	144.96	3.87
158	8.00	1300.	27	4241.	0.04890	-350.80	7.42
159	8.00	2759.	57	4264.	0.02830	-365.10	9.88
160	8.00	700.	14	4405.	0.06270	-158.42	1.05
161	8.00	1825.	38	4231.	0.07080	-229.33	6.45
162	8.00	1485.	31	4220.	0.02650	348.71	4.54
163	8.00	730.	15	4287.	0.12930	89.93	0.72
164	8.00	350.	7	4405.	0.11940	-196.57	1.53
165	8.00	1220.	25	4299.	0.14720	129.63	2.86
166	8.00	975.	20	4294.	0.15140	108.04	1.64
167	8.00	1600.	33	4271.	0.16070	98.62	2.37
168	8.00	600.	12	4405.	0.06230	26.27	0.02
169	8.00	1450.	30	4258.	0.06590	49.67	0.22
170	8.00	550.	11	4405.	0.06690	43.06	0.06
171	8.00	750.	15	4405.	0.06800	30.67	0.05
172	8.00	1550.	32	4267.	0.06780	23.83	0.06
173	8.00	600.	12	4405.	0.03560	111.75	0.25
174	8.00	500.	10	4405.	0.04150	84.75	0.14

175	8.00	2025.	42	4247.	0.04230	35.61	0.10
176	8.00	225.	5	3964.	0.03550	-57.11	0.02
177	8.00	600.	12	4405.	0.05700	-49.14	0.08
178	8.00	1150.	24	4221.	0.03460	49.77	0.09
179	8.00	555.	11	4445.	0.03270	73.69	0.09
180	8.00	575.	12	4221.	0.03150	78.29	0.11
181	8.00	1185.	24	4350.	0.03330	53.04	0.11
182	8.00	600.	12	4405.	0.02920	131.33	0.29
183	8.00	1625.	34	4210.	0.03310	97.84	0.49
184	8.00	2125.	44	4254.	0.03140	100.57	0.64
185	8.00	650.	13	4405.	0.02840	192.37	0.65
186	8.00	1975.	41	4243.	0.03100	105.30	0.64
187	8.00	830.	17	4301.	0.14470	95.68	1.04
188	8.00	1600.	33	4271.	0.13870	123.43	3.21
189	8.00	525.	11	4204.	0.17270	46.95	0.19
190	8.00	450.	9	4405.	0.17020	53.48	0.21
191	8.00	500.	10	4405.	0.20450	30.36	0.09
192	8.00	600.	12	4405.	0.05760	29.38	0.03
193	8.00	425.	9	4160.	0.05760	29.38	0.02
194	8.00	450.	9	4405.	0.05940	18.71	0.01
195	8.00	625.	13	4235.	0.05940	18.71	0.01
196	8.00	2350.	48	4313.	0.06290	1.41	0.00
197	8.00	2500.	52	4235.	0.03600	14.09	0.02
198	8.00	1950.	40	4294.	0.13040	-141.41	4.83
199	8.00	730.	15	4287.	0.12650	100.00	0.88
200	8.00	1075.	22	4304.	0.19030	17.47	0.06
201	8.00	400.	8	4405.	0.03520	17.02	0.00
202	6.00	750.	15	4405.	0.15030	84.91	3.25
203	6.00	235.	5	4140.	0.03010	-27.37	0.02
204	6.00	825.	17	4275.	0.14030	-27.37	0.35
205	6.00	1890.	38	4381.	0.13870	90.56	8.60
206	6.00	1700.	34	4405.	0.15360	-79.78	6.65
207	6.00	825.	17	4275.	0.15080	89.37	3.97
208	6.00	2025.	41	4351.	0.14910	89.65	9.70
209	6.00	1525.	31	4334.	0.18890	-32.55	1.22
210	6.00	1575.	32	4336.	0.16010	59.81	3.61
211	6.00	500.	10	4405.	0.14830	89.11	2.35
212	6.00	1300.	26	4405.	0.15230	71.44	4.04
213	6.00	1190.	24	4368.	0.13860	78.69	4.08
214	6.00	238.	5	4193.	0.13860	0.00	0.00
215	6.00	880.	18	4307.	0.14240	67.39	2.28
216	6.00	500.	10	4405.	0.14930	50.10	0.75
217	6.00	720.	15	4228.	0.23630	-0.01	0.00
218	6.00	465.	9	4551.	0.14620	-52.50	0.75
219	6.00	1947.	39	4398.	0.14030	70.30	5.40
220	6.00	200.	4	4405.	0.14200	0.00	0.00
221	6.00	1250.	25	4405.	0.14200	29.50	0.62
222	6.00	700.	14	4405.	0.12340	120.17	4.99
223	6.00	1960.	40	4317.	0.16610	46.94	2.87
224	6.00	1025.	21	4300.	0.19350	-1.77	0.00
225	6.00	850.	17	4405.	0.13810	149.68	10.52
226	6.00	875.	18	4282.	0.18230	30.11	0.58
227	6.00	625.	13	4235.	0.18110	-8.89	0.04
228	6.00	400.	8	4405.	0.14760	99.32	2.33
229	6.00	1175.	24	4313.	0.14080	121.01	9.69
230	6.00	550.	11	4405.	0.14670	113.96	4.19
231	6.00	570.	12	4184.	0.13330	-170.14	8.79

232	6.00	450.	9	4405.	0.14290	-54.90	0.78
233	6.00	1145.	23	4385.	0.15080	-90.55	5.66
234	6.00	525.	11	4204.	0.16010	-52.61	0.93
235	6.00	725.	15	4258.	0.13250	-58.16	1.30
236	6.00	2400.	49	4315.	0.14550	-37.76	1.99
237	6.00	400.	8	4405.	0.13510	-64.76	0.91
238	6.00	1880.	38	4358.	0.13670	58.79	3.55
239	6.00	1100.	22	4405.	0.17040	35.64	0.95
240	6.00	550.	11	4405.	0.14660	35.64	0.41
241	6.00	650.	13	4405.	0.12320	35.64	0.41
242	6.00	744.	15	4369.	0.14000	35.64	0.53
243	6.00	500.	10	4405.	0.16550	-40.65	0.55
244	6.00	925.	19	4289.	0.15920	-49.98	1.47
245	6.00	2540.	51	4387.	0.17260	-28.97	1.47
246	6.00	1075.	22	4304.	0.13800	138.12	11.32
247	6.00	625.	13	4235.	0.19750	25.08	0.31
248	6.00	750.	15	4405.	0.15900	-44.43	0.94
249	6.00	800.	16	4405.	0.13840	-133.81	7.93
250	6.00	1550.	31	4405.	0.18050	17.63	0.35
251	6.00	1375.	28	4326.	0.15770	61.21	3.25
252	6.00	1475.	30	4331.	0.15500	72.08	4.75
253	6.00	1200.	24	4405.	0.15850	-64.98	3.21
254	6.00	1450.	29	4405.	0.14270	-111.99	10.38
255	6.00	1725.	35	4342.	0.16140	64.65	4.65
256	6.00	825.	17	4275.	0.18080	43.08	1.11
257	6.00	1175.	24	4313.	0.22250	-22.87	0.55
258	6.00	1500.	30	4405.	0.22250	-15.84	0.34
259	6.00	725.	15	4258.	0.16880	17.59	0.15
260	6.00	1280.	26	4337.	0.16570	-28.69	0.70
261	6.00	650.	13	4405.	0.14540	-74.08	2.07
262	6.00	1225.	25	4317.	0.20770	36.20	1.33
263	6.00	1825.	37	4345.	0.14160	65.55	4.44
264	6.00	1250.	25	4405.	0.16950	58.94	2.94
265	6.00	3665.	74	4363.	0.17310	-13.56	0.47
266	6.00	850.	17	4405.	0.16330	41.06	0.94
267	6.00	1950.	39	4405.	0.19490	13.26	0.27
268	6.00	2425.	49	4360.	0.14140	74.11	7.53
269	6.00	1750.	35	4405.	0.07970	54.55	1.66
270	6.00	1625.	33	4338.	0.07970	49.65	1.28
271	6.00	2025.	41	4351.	0.04220	117.03	4.68
272	6.00	975.	20	4294.	0.04420	70.88	0.87
273	6.00	1975.	40	4350.	0.05720	87.85	3.49
274	6.00	750.	15	4405.	0.08220	19.91	0.10
275	6.00	1830.	37	4357.	0.08760	-1.89	0.00
276	6.00	2075.	42	4352.	0.09430	-17.69	0.24
277	6.00	3250.	66	4338.	0.07890	107.59	11.87
278	6.00	1000.	20	4405.	0.05410	219.37	10.41
279	6.00	1050.	21	4405.	0.04860	139.56	3.97
280	6.00	2200.	45	4307.	0.05410	113.48	6.13
281	6.00	1100.	22	4405.	0.24640	10.66	0.12
282	6.00	800.	16	4405.	0.13680	79.52	2.77
283	6.00	1825.	37	4345.	0.15020	55.42	3.37
284	6.00	2630.	53	4371.	0.14580	74.20	8.44
285	6.00	2665.	54	4348.	0.14090	85.88	11.08
286	6.00	1825.	37	4345.	0.05150	99.29	3.70
287	6.00	4200.	85	4353.	0.05350	148.88	19.92
288	6.00	650.	13	4405.	0.13660	-192.02	13.09

289	6.00	225.	5	3964.	0.03130	101.84	0.29
290	6.00	600.	12	4405.	0.03240	89.04	0.62
291	6.00	1425.	29	4329.	0.06790	-8.70	0.03
292	6.00	1275.	26	4320.	0.03470	8.70	0.01
293	6.00	575.	12	4221.	0.05540	81.46	0.85
294	6.00	3250.	66	4338.	0.05820	2.82	0.01
295	6.00	1625.	33	4338.	0.05580	15.39	0.09
296	6.00	1825.	37	4345.	0.05690	-2.40	0.00
297	6.00	625.	13	4235.	0.08060	1.34	0.00
298	6.00	1400.	28	4405.	0.08850	-1.07	0.00
299	6.00	2225.	45	4356.	0.18950	7.50	0.09
300	6.00	725.	15	4258.	0.18030	10.50	0.06
301	6.00	1815.	37	4321.	0.17880	29.11	1.10
302	6.00	2225.	45	4356.	0.19180	-11.11	0.21
303	6.00	2575.	52	4362.	0.19280	25.89	1.33
304	6.00	925.	19	4289.	0.31270	-2.50	0.01
305	6.00	3025.	61	4368.	0.20390	21.87	1.18
306	6.00	4160.	84	4363.	0.19510	20.63	1.38
307	6.00	625.	13	4235.	0.18290	19.29	0.17
308	6.00	1350.	27	4405.	0.19380	13.88	0.20
309	6.00	1900.	38	4405.	0.13360	105.84	11.37
310	6.00	1465.	30	4302.	0.13570	76.08	4.60
311	6.00	1420.	29	4313.	0.13530	77.39	4.60
312	6.00	1125.	23	4309.	0.13190	98.66	5.78
313	6.00	468.	9	4581.	0.17200	-18.82	0.11
314	6.00	1070.	22	4284.	0.13520	99.86	5.77
315	6.00	300.	6	4405.	0.15150	-24.34	0.11
316	6.00	400.	8	4405.	0.14820	67.20	1.07
317	6.00	475.	10	4184.	0.13880	66.82	1.18
318	6.00	450.	9	4405.	0.18850	-1.82	0.00
319	6.00	850.	17	4405.	0.14460	53.00	1.38
320	6.00	700.	14	4405.	0.14770	64.52	1.72
321	6.00	1335.	27	4356.	0.13420	86.13	5.31
322	6.00	1340.	27	4372.	0.14400	81.29	5.10
323	6.00	2119.	43	4341.	0.13430	93.12	9.87
324	6.00	675.	14	4247.	0.04910	-67.76	0.61
325	6.00	500.	10	4405.	0.18610	18.38	0.13
326	6.00	1300.	26	4405.	0.14170	99.67	7.32
327	6.00	2050.	41	4405.	0.06480	42.41	0.96
328	6.00	2840.	57	4389.	0.11320	10.40	0.14
329	6.00	1875.	38	4347.	0.06000	35.18	0.56
330	6.00	750.	15	4405.	0.07690	110.35	2.81
331	6.00	325.	7	4090.	0.07580	52.00	0.27
332	6.00	600.	12	4405.	0.07910	4.13	0.00
333	6.00	600.	12	4405.	0.07100	-23.40	0.09
334	6.00	1050.	21	4405.	0.09040	77.86	2.30
335	6.00	400.	8	4405.	0.06760	48.46	0.25
336	6.00	800.	16	4405.	0.09930	-56.18	1.00
337	6.00	775.	16	4267.	0.08740	-34.05	0.31
338	6.00	250.	5	4405.	0.08100	67.55	0.37
339	6.00	1425.	29	4329.	0.09360	-24.55	0.32
340	6.00	275.	6	4038.	0.08980	-0.64	0.00
341	6.00	1345.	27	4388.	0.10010	-35.82	0.69
342	6.00	500.	10	4405.	0.08470	35.19	0.21
343	6.00	1475.	30	4331.	0.06520	-9.40	0.03
344	6.00	330.	7	4153.	0.05790	21.08	0.03
345	6.00	450.	9	4405.	0.05720	16.98	0.03

346	6.00	375.	8	4129.	0.03500	-77.73	0.32
347	6.00	675.	14	4247.	0.06410	53.49	0.50
348	6.00	2725.	55	4365.	0.05930	8.43	0.05
349	6.00	500.	10	4405.	0.13610	-41.96	0.48
350	6.00	775.	16	4267.	0.04130	39.92	0.20
351	6.00	495.	10	4361.	0.14900	23.03	0.16
352	4.00	1110.	22	4445.	0.14900	23.03	2.66
353	6.00	425.	9	4160.	0.20810	-27.92	0.28
354	6.00	625.	13	4235.	0.03080	16.33	0.02
355	6.00	1375.	28	4326.	0.08840	-11.39	0.06
356	6.00	600.	12	4405.	0.07170	27.27	0.13
357	6.00	1375.	28	4326.	0.10260	0.38	0.00
358	6.00	600.	12	4405.	0.07730	-15.65	0.05
359	6.00	705.	14	4436.	0.05770	46.54	0.35
360	6.00	740.	15	4346.	0.05880	62.91	0.69
361	6.00	775.	16	4267.	0.06210	23.12	0.10
362	6.00	650.	13	4405.	0.06730	14.12	0.03
363	6.00	350.	7	4405.	0.03740	-0.01	0.00
364	20.00	100.	2	4405.	0.02010	-4715.49	0.43
365	24.00	100.	2	4405.	0.02660	738.03	0.01
366	20.00	544.	12	3994.	0.03670	2543.84	1.26
367	20.00	500.	11	4004.	0.03720	2635.52	1.26
368	30.00	212.	5	3735.	0.01820	11436.47	0.65
369	24.00	544.	13	3686.	0.01150	6257.11	0.96
370	12.00	100.	2	4405.	0.03000	440.60	0.07
371	12.00	1371.	29	4165.	0.07970	104.20	0.15

CONNECTIVITY

LINE	U/S NODE	D/S NODE
1	1	2
2	2	3
3	3	4
4	4	5
5	5	6
6	6	7
7	7	8
8	8	9
9	9	10
10	10	11
11	11	12
12	12	13
13	13	14
14	14	15
15	15	16
16	16	17
17	17	18
18	18	19
19	19	20
20	20	21
21	21	22
22	22	23
23	21	24
24	24	25
25	25	26
26	26	27
27	27	28
28	28	29
29	29	30
30	30	31
31	31	4
32	32	33
33	33	34
34	34	35
35	35	36
36	36	37
37	37	30
38	30	38
39	38	39
40	39	40
41	40	41
42	41	42
43	42	43
44	43	10
45	10	44
46	44	45
47	41	46
48	46	8
49	35	47
50	47	48
51	48	49
52	22	50
53	50	51

54	51	52
55	52	53
56	20	54
57	54	55
58	55	56
59	12	57
60	57	58
61	58	59
62	28	60
63	28	61
64	61	62
65	62	63
66	63	64
67	61	65
68	51	66
69	66	67
70	66	68
71	68	69
72	21	70
73	70	50
74	12	71
75	71	72
76	72	73
77	73	74
78	75	76
79	26	77
80	26	78
81	19	79
82	80	81
83	50	69
84	50	51
85	51	67
86	24	18
87	25	17
88	17	82
89	82	83
90	83	84
91	84	85
92	85	86
93	86	87
94	87	88
95	88	89
96	89	90
97	90	91
98	91	92
99	92	93
100	93	94
101	94	95
102	95	56
103	56	96
104	96	97
105	97	98
106	98	99
107	99	82
108	84	100
109	100	101
110	101	102

111	102	103
112	103	104
113	104	13
114	105	106
115	106	107
116	107	108
117	108	109
118	55	110
119	27	111
120	105	112
121	112	33
122	33	3
123	33	113
124	113	114
125	115	116
126	116	117
127	117	77
128	78	118
129	118	119
130	119	120
131	120	40
132	40	121
133	121	122
134	122	6
135	6	123
136	123	124
137	124	125
138	125	126
139	126	127
140	125	108
141	128	15
142	16	129
143	129	130
144	130	131
145	131	132
146	132	133
147	131	134
148	58	135
149	135	136
150	136	137
151	102	138
152	139	140
153	140	88
154	140	85
155	141	142
156	24	143
157	52	80
158	54	144
159	144	19
160	53	145
161	145	110
162	110	146
163	146	147
164	147	95
165	146	148
166	148	149
167	149	81

168	81	150
169	150	151
170	151	152
171	152	153
172	153	154
173	94	155
174	155	156
175	156	158
176	158	157
177	157	156
178	96	159
179	96	159
180	159	161
181	159	161
182	161	162
183	97	163
184	163	164
185	164	90
186	162	164
187	92	165
188	165	166
189	166	167
190	167	168
191	168	169
192	169	170
193	170	171
194	171	172
195	172	173
196	173	174
197	174	175
198	175	92
199	175	176
200	57	58
201	162	163
202	109	126
203	159	160
204	160	56
205	112	177
206	4	177
207	106	177
208	177	178
209	5	178
210	107	178
211	178	124
212	5	123
213	113	179
214	179	180
215	179	181
216	181	182
217	181	114
218	182	114
219	182	183
220	183	184
221	34	183
222	34	185
223	185	189
224	185	186

225	114	186
226	186	187
227	187	188
228	188	189
229	4	188
230	189	190
231	190	37
232	190	191
233	191	39
234	39	38
235	38	30
236	30	195
237	195	48
238	30	29
239	191	192
240	192	193
241	193	194
242	194	121
243	5	196
244	196	31
245	196	31
246	123	197
247	197	198
248	198	199
249	199	7
250	197	201
251	198	200
252	199	205
253	200	201
254	201	127
255	201	202
256	200	203
257	203	205
258	202	203
259	204	205
260	203	204
261	204	45
262	202	206
263	45	206
264	206	207
265	207	59
266	58	208
267	208	137
268	43	13
269	73	209
270	74	209
271	210	119
272	120	210
273	42	210
274	118	211
275	211	212
276	212	128
277	43	128
278	74	138
279	15	130
280	14	213
281	213	104

282	213	133
283	133	103
284	132	101
285	134	100
286	129	134
287	100	214
288	214	86
289	83	215
290	215	139
291	139	216
292	99	216
293	84	139
294	89	142
295	87	141
296	86	217
297	141	217
298	217	142
299	219	218
300	220	219
301	64	220
302	221	220
303	63	221
304	222	221
305	223	222
306	60	222
307	62	223
308	60	223
309	60	224
310	111	225
311	111	225
312	225	227
313	224	225
314	224	226
315	226	227
316	226	229
317	227	228
318	229	228
319	228	116
320	229	115
321	116	231
322	115	230
323	26	143
324	143	231
325	231	230
326	230	21
327	145	148
328	91	232
329	166	174
330	146	238
331	148	235
332	149	234
333	150	233
334	234	233
335	233	244
336	234	236
337	234	75
338	235	236

339	235	238
340	236	237
341	237	238
342	237	241
343	239	238
344	238	239
345	239	157
346	240	158
347	240	241
348	240	246
349	75	241
350	241	243
351	76	242
352	242	167
353	76	243
354	154	169
355	151	244
356	244	245
357	152	245
358	153	245
359	246	165
360	93	246
361	168	247
362	247	171
363	82	248
364	2	32
365	1	105
366	249	1
367	249	1
368	250	249
369	249	32
370	67	251
371	209	252

APPENDIX G
SYSTEM CHARACTERISTICS FOR MODEL CITY A

STEADY STATE RESULTS
LARGE SYSTEM HYDRANT @ NODE 181 CLOSED

HYDRAULIC STEADY STATE CALCULATION
OUTPUT CREATED BY NETWORK VERSION 00.01

PROJECT: AWWA LARGE
RUN : NODE 181 CL

****
NODE STEADY STATE DATA
****

NODE	TYPE	ID	AQUE INVERT (FT)	HGL (FT)	SPR-LN PRESS (FT)	FLOW PAST (GPM)	DISCHG (GPM)
1	P	23	603.0	883.6	280.6	0.0	105913.6
2	V	27	650.0	871.6	221.6	36107.8	222.0
3	J	33	620.0	862.9	242.9	34051.0	0.0
4	V	35	649.0	858.6	209.6	0.0	201.0
5	V	36	670.0	857.7	187.7	14128.5	361.0
6	V	38	845.0	855.5	10.5	20841.6	146.0
7	R	39 H-P RES	855.0	855.0	0.0	20695.6	2573.1
8	V	50	705.0	853.7	148.7	18122.4	625.0
9	V	79	763.0	851.4	88.4	18462.7	403.0
10	V	91	710.0	849.2	139.2	16830.7	188.0
11	V	93	710.0	845.0	135.0	16642.7	285.0
12	V	98	710.0	826.9	116.9	15446.8	361.0
13	V	100	712.0	825.4	113.4	12891.4	797.0
14	J	101	712.0	824.6	112.6	10315.7	0.0
15	V	107	718.0	821.9	103.9	10612.9	785.0
16	V	109	718.0	821.3	103.3	6020.1	493.0
17	V	113	689.0	820.0	131.0	6065.2	319.0
18	J	157	630.0	806.2	176.2	21658.9	0.0
19	V	158	636.0	797.7	161.7	15422.0	181.0
20	V	159	638.0	796.4	158.4	15578.3	5660.0
21	J	160	640.0	766.0	126.0	23052.9	0.0
22	J	167 LOW ELEV	770.0	777.6	7.6	-4740.7	0.0
23	R	301 LOOKOUTR	785.0	785.1	0.1	-4740.7	12977.3
24	J	912	686.0	796.4	110.4	-17718.0	0.0
25	V	190	640.0	796.8	156.8	-16339.9	104.0
26	J	302	630.0	807.1	177.1	-16443.9	0.0
27	V	197	620.0	807.6	187.6	-3509.8	243.0
28	V	303	630.0	810.0	180.0	-6988.8	528.0
29	V	198	634.0	811.5	177.5	-7516.8	549.0
30	V	315	637.0	813.6	176.6	-7913.4	104.0
31	V	219	640.0	820.2	180.2	-9691.1	465.0
32	V	230	640.0	820.2	180.2	-872.2	451.0
33	V	242	625.0	820.3	195.3	-1323.2	917.0
34	V	247	623.0	820.5	197.5	-2240.2	715.0
35	J	248	663.0	820.7	157.7	-2955.2	0.0
36	J	304	727.0	821.0	94.0	-2955.2	0.0

37	V	253		744.0	821.4	77.4	-2955.2	354.0
38	V	920		686.0	820.4	134.4	3250.0	1431.0
39	V	921		709.0	820.4	111.4	1819.0	215.0
40	J	922		720.0	820.3	100.3	1604.0	0.0
41	V	923		735.0	820.2	85.2	1604.0	1604.0
42	V	34		620.0	860.0	240.0	346.2	243.0
43	J	28		628.0	871.8	243.8	16706.7	0.0
44	V	42		656.0	864.1	208.1	8295.4	236.0
45	V	55		720.0	854.2	134.2	8857.0	403.0
46	J	69		711.0	847.0	136.0	8628.5	0.0
47	V	70		715.0	845.6	130.6	3642.8	444.0
48	J	910		764.0	820.5	56.5	-6161.0	0.0
49	V	140		772.0	821.5	49.5	-4933.1	313.0
50	V	145		776.0	818.8	42.8	9401.5	229.0
51	V	144		794.0	817.9	23.9	9172.5	188.0
52	**R**	**143**	**ANDRSNR**	**817.0**	**817.0**	**0.0**	**8984.5**	**11805.0**
53	**V**	**148**		**789.0**	**817.0**	**28.0**	**-746.7**	**528.0**
54	V	169		752.0	817.1	65.1	-1092.2	583.0
55	V	186		727.0	817.2	90.2	-638.6	1167.0
56	V	185		720.0	822.8	102.8	-10238.9	1222.0
57	V	184		735.0	828.3	93.3	-11427.2	903.0
58	V	203		746.0	834.3	88.3	-23512.9	500.0
59	V	214		785.0	841.2	56.2	-25195.8	417.0
60	V	213		776.0	835.6	59.6	24497.8	479.0
61	V	212		694.0	830.3	136.3	24018.8	188.0
62	V	310		731.0	829.7	98.7	8322.2	167.0
63	J	311		735.0	829.1	94.1	8155.2	0.0
64	V	312		667.0	828.5	161.5	8155.2	194.0
65	J	305		691.0	827.9	136.9	7961.2	0.0
66	V	313		719.0	825.2	106.2	7961.2	1083.0
67	V	252		741.0	823.2	82.2	319.0	319.0
68	V	183		679.0	828.2	149.2	3366.9	618.0
69	V	182		663.0	828.1	165.1	2949.5	729.0
70	V	179		680.0	827.9	147.9	6043.7	625.0
71	V	308		667.0	827.8	160.8	5418.7	1396.0
72	V	178		710.0	827.7	117.7	3151.8	1181.0
73	V	207		690.0	828.0	138.0	-5928.9	1924.0
74	V	209		702.0	829.3	127.3	10373.9	1146.0
75	V	208		700.0	828.6	128.6	1375.0	1375.0
76	J	29		610.0	871.5	261.5	36618.2	0.0
77	V	43		623.0	863.7	240.7	28408.5	160.0
78	V	56		674.0	852.8	178.8	25826.0	201.0
79	V	68		718.0	843.1	125.1	30610.7	250.0
80	V	67		720.0	841.4	121.4	30360.7	222.0
81	V	122		725.0	830.0	105.0	29439.8	361.0
82	V	139		730.0	827.3	97.3	1157.1	250.0
83	V	138		764.0	827.3	63.3	907.1	250.0
84	V	137		710.0	827.3	117.3	657.1	500.0
85	V	172		694.0	827.6	133.6	2188.4	1542.0
86	V	136		688.0	825.7	137.7	5784.6	486.0
87	V	135		633.0	824.2	191.2	5821.8	292.0
88	V	134		644.0	824.2	180.2	-924.8	333.0
89	V	132		700.0	824.4	124.4	-2900.9	2285.0
90	V	131		760.0	824.6	64.6	-4734.1	1924.0
91	J	130		784.0	825.2	41.2	-6658.1	0.0
92	J	129		636.0	825.5	189.5	-6658.1	0.0
93	J	902		643.0	825.4	182.4	8553.0	0.0

94	V	46	660.0	761.9	101.9	8553.0	111.0
95	V	17	584.0	737.3	153.3	4640.0	104.0
96	V	15	590.0	727.9	137.9	4536.0	313.0
97	V	8	580.0	723.0	143.0	3257.0	222.0
98	V	12	570.0	717.3	147.3	3035.0	5035.0
99	V	30	600.0	845.6	245.6	8802.0	986.0
100	J	22	603.0	793.3	190.3	7412.3	0.0
101	J	904	600.0	770.2	170.2	5603.7	0.0
102	R	16 CATOOSA	595.0	725.1	130.1	8828.4	11919.3
103	V	5	583.0	722.5	139.5	2236.0	236.0
104	J	4	624.0	719.5	95.5	0.0	0.0
105	J	32	666.0	877.6	211.6	24892.1	0.0
106	V	41	706.0	871.4	165.4	12468.7	222.0
107	V	53	706.0	865.1	159.1	12246.7	632.0
108	V	54	680.0	862.0	182.0	10085.1	201.0
109	V	71	711.0	853.8	142.8	4959.6	667.0
110	V	192	690.0	867.7	177.7	11033.5	209.0
111	V	52	688.0	860.6	172.6	11010.4	403.0
112	V	75	756.0	851.0	95.0	9276.5	146.0
113	J	77	730.0	845.6	115.6	4908.4	0.0
114	V	76	722.0	844.5	122.5	8526.6	201.0
115	V	117	700.0	833.1	133.1	8325.6	938.0
116	J	905	707.0	820.1	113.1	6270.9	0.0
117	V	151	713.0	817.0	104.0	3699.2	1201.0
118	V	152	741.0	816.7	75.7	2498.2	417.0
119	V	168	757.0	814.3	57.3	2451.9	646.0
120	V	188	740.0	813.5	73.5	2004.0	389.0
121	V	187	700.0	813.0	113.0	10629.1	514.0
122	V	199	705.0	813.4	108.4	-1696.7	174.0
123	V	21	620.0	827.2	207.2	1399.0	160.0
124	V	7	600.0	767.4	167.4	1306.0	528.0
125	V	6	630.0	766.8	136.8	778.0	778.0
126	V	26	630.0	832.6	202.6	1818.3	110.0
127	V	20	640.0	827.3	187.3	529.9	319.0
128	V	19	666.0	826.9	160.9	210.9	604.0
129	V	307	650.0	826.7	176.7	778.0	778.0
130	V	24	690.0	826.6	136.6	0.0	257.0
131	V	51	700.0	859.6	159.6	486.0	486.0
132	V	40	693.0	868.0	175.0	250.0	250.0
133	V	80	712.0	851.4	139.4	1229.0	111.0
134	V	81	688.0	850.9	162.9	1118.0	1118.0
135	V	92	710.0	823.2	113.2	6454.6	403.0
136	V	89	710.0	818.3	108.3	5709.4	333.0
137	V	88	638.0	817.9	179.9	806.0	806.0
138	J	115	695.0	816.3	121.3	4570.4	0.0
139	V	114	700.0	813.4	113.4	-5999.8	541.0
140	V	263	700.0	822.0	122.0	1424.2	313.0
141	J	99	710.0	821.0	111.0	1111.2	0.0
142	V	108	700.0	815.3	115.3	396.0	396.0
143	V	94	715.0	828.0	113.0	910.8	174.0
144	V	112	655.0	820.0	165.0	768.1	174.0
145	V	156	670.0	817.6	147.6	1342.0	103.0
146	V	155	688.0	816.5	128.5	1239.0	146.0
147	V	154	694.0	814.5	120.5	1093.0	278.0
148	V	153	700.0	813.7	113.7	815.0	486.0
149	V	189	640.0	810.4	170.4	329.0	728.0
150	J	913	638.0	811.8	173.8	0.0	0.0

151	J	162		633.0	797.7	164.7	-337.3	0.0
152	V	164		635.0	797.8	162.8	-337.3	215.0
153	V	165		635.0	798.2	163.2	-552.3	1319.0
154	V	166		633.0	796.4	163.4	312.0	208.0
155	V	190		640.0	796.1	156.1	104.0	104.0
156	V	163		635.0	802.1	167.1	111.0	111.0
157	V	222		681.0	806.4	125.4	0.0	263.0
158	V	221		752.0	805.5	53.5	1618.0	1618.0
159	V	191		729.0	806.2	77.2	0.0	243.0
160	V	193		681.0	806.2	125.2	-243.0	681.0
161	V	194		718.0	806.3	88.3	-924.0	431.0
162	J	195		721.0	806.5	85.5	-1355.0	0.0
163	J	196		703.0	806.7	103.7	0.0	0.0
164	V	218		734.0	860.6	126.6	992.0	313.0
165	V	74		751.0	851.0	100.0	2486.3	201.0
166	V	78		740.0	836.2	96.2	250.0	250.0
167	V	95		713.0	825.9	112.9	1448.2	132.0
168	V	96		713.0	825.8	112.8	1316.2	174.0
169	V	97		715.0	825.8	110.8	2139.4	111.0
170	V	262		730.0	825.3	95.3	2725.0	313.0
171	J	103		711.0	824.8	113.8	1197.2	0.0
172	V	102		730.0	824.7	94.7	2412.0	278.0
173	V	106		722.0	823.4	101.4	116.2	278.0
174	V	104		707.0	825.8	118.8	1116.7	243.0
175	J	105		703.0	824.0	121.0	0.0	0.0
176	V	110		720.0	821.7	101.7	326.0	326.0
177	J	73		730.0	849.0	119.0	2285.3	0.0
178	J	306		685.0	813.5	128.5	-1870.7	0.0
179	V	200		685.0	814.7	129.7	-1308.6	252.0
180	V	201		672.0	823.2	151.2	-2141.3	500.0
181	V	72		629.0	847.7	218.7	1598.0	0.0
182	V	119		730.0	830.3	100.3	1598.0	486.0
183	J	906		730.0	820.1	90.1	0.0	0.0
184	V	149		730.0	815.1	85.1	563.0	563.0
185	V	317		740.0	847.3	107.3	2815.1	243.0
186	V	120		760.0	833.0	73.0	2572.1	340.0
187	V	911		749.0	829.4	80.4	931.0	931.0
188	V	121		748.0	822.4	74.4	1548.9	285.0
189	J	908		779.0	820.5	41.5	1263.9	0.0
190	V	146		810.0	818.2	8.2	132.0	132.0
191	J	909		768.0	820.5	52.5	0.0	0.0
192	V	141		785.0	818.4	33.4	4036.9	285.0
193	V	142		817.0	817.4	0.4	174.0	174.0
194	V	171		740.0	818.4	78.4	1091.4	785.0
195	V	170		743.0	817.5	74.5	417.0	417.0
196	V	217		711.0	824.7	113.7	-12020.9	1035.0
197	V	202		710.0	827.2	117.2	-331.3	271.0
198	V	216		729.0	832.5	103.5	708.0	708.0
199	R	215	SHER RES	850.0	848.0	-2.0	-64476.8	-64476.8
200	V	204		750.0	828.8	78.8	354.5	465.0
201	V	205		674.0	829.0	155.0	-3933.7	694.0
202	V	211		690.0	829.8	139.8	507.0	507.0
203	V	175		670.0	816.6	146.6	870.9	1563.0
204	V	176		690.0	827.4	137.4	2403.0	2403.0
205	V	44		660.0	843.1	183.1	3172.2	354.0
206	V	49		644.0	802.5	158.5	2818.2	1104.0
207	V	58		626.0	794.8	168.8	1714.2	264.0

208	V	57		710.0	792.8	82.8	706.2	500.0
209	V	59		620.0	792.7	172.7	243.0	243.0
210	J	903		688.0	794.8	106.8	464.2	0.0
211	V	60		722.0	774.2	52.2	3052.0	298.0
212	J	48		730.0	771.3	41.3	2754.0	0.0
213	V	47		660.0	766.6	106.6	2754.0	146.0
214	V	63		617.0	761.4	144.4	1083.0	1083.0
215	V	125		624.0	812.7	188.7	-6502.8	882.0
216	V	135		633.0	829.6	196.6	-7384.8	292.0
217	V	66		650.0	829.6	179.6	174.0	174.0
218	V	65		670.0	792.8	122.8	243.0	243.0
219	V	64		700.0	832.8	132.8	698.9	208.0
220	V	123		700.0	830.7	130.7	490.9	250.0
221	V	124		720.0	829.9	109.9	9237.2	354.0
222	V	264		615.0	829.8	214.8	750.0	750.0
223	V	265		623.0	793.4	170.4	-503.5	153.0
224	V	180		600.0	793.6	193.6	1000.0	1000.0
225	V	3		590.0	791.4	201.4	2312.0	1069.0
226	V	174		641.0	759.6	118.6	1243.0	1243.0
227	V	182		663.0	824.1	161.1	951.0	729.0
228	V	25		640.0	824.1	184.1	222.0	222.0
229	J	309		754.0	827.6	73.6	15211.1	0.0
230	J	901		709.0	826.6	117.6	0.0	0.0
231	R	161	WTULSAR	656.0	750.2	94.2	27793.7	27793.7

****
PIPELINE STEADY STATE DATA
****

LINE	DIAMETER (IN)	LENGTH (FT)	SECTIONS	WAVE SP. (FT/S)	(DARCY)	DISCHARGE (GPM)	HEAD LOSS (FT)
1	48.00	7500.	80	3415.	0.01000	36107.84	11.93
2	48.00	6200.	66	3422.	0.01000	34050.95	8.77
3	30.00	200.	2	3643.	0.01100	32232.61	2.92
4	48.00	3700.	40	3370.	0.01000	14128.54	0.90
5	48.00	4200.	45	3400.	0.01000	20841.58	2.23
6	48.00	900.	10	3279.	0.01000	20695.58	0.47
7	48.00	3300.	35	3435.	0.01000	18122.43	1.32
8	48.00	5500.	59	3396.	0.01000	18462.66	2.29
9	42.00	2200.	22	3643.	0.01480	16830.66	2.19
10	30.00	440.	4	4007.	0.02670	16642.66	4.16
11	30.00	2230.	22	3693.	0.02670	15446.84	18.18
12	30.00	410.	4	3734.	0.01650	12891.40	1.44
13	30.00	380.	4	3461.	0.01650	10315.74	0.85
14	30.00	1140.	11	3775.	0.01650	10612.93	2.71
15	30.00	700.	7	3643.	0.01650	6020.13	0.54
16	30.00	1700.	17	3643.	0.01650	6065.18	1.32
17	36.00	5500.	58	3455.	0.01040	21658.88	13.80
18	30.00	1700.	17	3643.	0.01650	15422.02	8.54
19	30.00	250.	2	4554.	0.01650	15578.28	1.28
20	36.00	6010.	60	3649.	0.01850	23052.93	30.39
21	24.00	5350.	53	3677.	0.02480	-4740.72	11.64
22	24.00	3440.	34	3686.	0.02480	-4740.72	7.49
23	36.00	4500.	45	3643.	0.01550	-17718.04	11.26
24	36.00	200.	2	3643.	0.01550	-16339.93	0.43

25	36.00	5280.	53	3629.	0.01550	-15610.89	10.26
26	36.00	5280.	53	3629.	0.01550	-3509.79	0.52
27	36.00	3320.	33	3665.	0.02890	-6988.79	2.41
28	36.00	1800.	18	3643.	0.02890	-7516.79	1.51
29	36.00	2560.	26	3587.	0.02520	-7913.43	2.08
30	36.00	8800.	88	3643.	0.01550	-9691.11	6.59
31	36.00	5280.	53	3629.	0.01550	-872.21	0.03
32	36.00	5280.	53	3629.	0.01550	-1323.21	0.07
33	36.00	6560.	66	3621.	0.01550	-2240.21	0.26
34	36.00	2600.	26	3643.	0.01550	-2806.04	0.16
35	12.00	2620.	24	3977.	0.02240	-149.17	0.16
36	36.00	5280.	53	3629.	0.01550	-2805.82	0.33
37	12.00	5280.	48	4007.	0.02250	-149.39	0.33
38	36.00	5280.	53	3629.	0.01550	-2955.21	0.37
39	36.00	17500.	184	3465.	0.01040	3250.00	0.99
40	36.00	2640.	28	3435.	0.01040	1819.00	0.05
41	36.00	5280.	55	3497.	0.01040	1604.00	0.07
42	36.00	5280.	55	3497.	0.01040	1604.00	0.07
43	36.00	5300.	53	3643.	0.01550	16706.72	11.79
44	30.00	5280.	52	3699.	0.01650	8295.39	7.67
45	30.00	5980.	59	3692.	0.01650	8857.00	9.90
46	30.00	4600.	46	3643.	0.01650	8628.54	7.23
47	24.00	1500.	15	3643.	0.01780	3642.84	1.38
48	24.00	7200.	71	3694.	0.01780	7083.78	25.11
49	24.00	600.	6	3643.	0.01780	-4933.12	1.01
50	36.00	3840.	38	3681.	0.01550	9401.50	2.71
51	36.00	1320.	13	3699.	0.01550	9172.50	0.89
52	36.00	1360.	14	3539.	0.01550	8984.50	0.88
53	36.00	1350.	14	3513.	0.01550	-746.71	0.01
54	36.00	5280.	53	3629.	0.01550	-1092.23	0.05
55	24.00	5280.	52	3699.	0.01780	-638.55	0.15
56	36.00	4750.	48	3605.	0.02170	-10238.91	5.56
57	36.00	5280.	53	3629.	0.01550	-11427.23	5.50
58	48.00	5280.	54	3562.	0.01680	-23512.92	5.98
59	48.00	5280.	54	3562.	0.01680	-25195.79	6.87
60	48.00	5350.	55	3544.	0.01410	24497.84	5.52
61	48.00	5350.	55	3544.	0.01410	24018.84	5.31
62	48.00	5285.	54	3565.	0.01410	8322.21	0.63
63	48.00	5290.	54	3569.	0.01410	8155.21	0.61
64	48.00	5260.	54	3549.	0.01410	8155.21	0.60
65	48.00	5285.	54	3565.	0.01410	7961.21	0.58
66	36.00	5300.	53	3643.	0.01550	7961.21	2.68
67	36.00	5300.	53	3643.	0.01550	6878.21	2.00
68	36.00	500.	5	3643.	0.01550	8411.33	0.28
69	48.00	5000.	51	3572.	0.01410	3366.86	0.10
70	48.00	6200.	64	3529.	0.01410	2949.54	0.09
71	48.00	3570.	37	3515.	0.01410	6043.73	0.22
72	48.00	1280.	13	3587.	0.01410	5418.73	0.06
73	48.00	5320.	55	3524.	0.01410	3075.54	0.09
74	12.00	5320.	49	3955.	0.02240	76.25	0.09
75	48.00	5260.	54	3549.	0.01410	-5928.93	0.32
76	48.00	5260.	54	3549.	0.01410	-7852.93	0.56
77	48.00	5260.	54	3549.	0.01410	10373.93	0.97
78	48.00	5280.	54	3562.	0.01410	9005.09	0.74
79	12.00	5300.	48	4022.	0.02240	222.84	0.74
80	48.00	5430.	56	3532.	0.02290	28206.92	12.07
81	48.00	5900.	61	3524.	0.01410	27816.23	7.85

82	48.00	6000.	62	3525.	0.02290	25479.98	10.89
83	48.00	5280.	54	3562.	0.02290	25624.98	9.69
84	48.00	650.	7	3383.	0.02290	30360.69	1.67
85	48.00	4700.	48	3567.	0.02290	29439.82	11.38
86	48.00	3950.	41	3510.	0.01410	19841.64	2.68
87	48.00	3400.	35	3539.	0.01410	907.12	0.00
88	48.00	2200.	23	3485.	0.01410	657.12	0.00
89	48.00	5400.	56	3513.	0.01410	-5627.46	0.29
90	48.00	5280.	54	3562.	0.01680	-7815.83	0.66
91	36.00	5050.	51	3607.	0.01850	5784.58	1.61
92	36.00	4670.	47	3620.	0.01850	5821.78	1.51
93	36.00	1180.	12	3582.	0.01550	-924.82	0.01
94	36.00	2100.	21	3643.	0.01550	-18684.53	5.84
95	48.00	9270.	95	3555.	0.01410	-2900.88	0.13
96	48.00	4960.	51	3543.	0.01970	-4734.11	0.27
97	48.00	5280.	54	3562.	0.01970	-6658.11	0.56
98	60.00	9500.	98	3531.	0.01830	-6658.11	0.31
99	72.00	5400.	72	2732.	0.00900	8553.00	0.06
100	24.00	12500.	123	3702.	0.01780	8553.00	63.56
101	24.00	16450.	162	3699.	0.01780	4640.04	24.62
102	24.00	6580.	65	3688.	0.01780	4536.04	9.41
103	24.00	6650.	66	3671.	0.01780	3256.99	4.90
104	24.00	8880.	88	3676.	0.01780	3034.99	5.69
105	24.00	4820.	48	3658.	0.01780	8802.02	25.96
106	24.00	13670.	135	3689.	0.01780	7412.27	52.20
107	24.00	10600.	105	3678.	0.01780	5603.72	23.14
108	24.00	10500.	104	3678.	0.01780	7862.35	45.12
109	24.00	7500.	74	3692.	0.01780	2236.01	2.61
110	16.00	4900.	46	3881.	0.02040	-966.06	2.77
111	24.00	10840.	107	3691.	0.01780	2000.01	3.01
112	24.00	7920.	78	3699.	0.01780	2000.01	2.20
113	36.00	4500.	45	3643.	0.01550	12954.91	6.02
114	36.00	5300.	53	3643.	0.01550	11937.21	6.02
115	36.00	4000.	40	3643.	0.01550	13858.58	6.12
116	36.00	5280.	53	3629.	0.01550	12246.74	6.31
117	30.00	1310.	13	3671.	0.01650	10605.67	3.11
118	30.00	3970.	39	3708.	0.01650	9884.12	8.19
119	24.00	4880.	48	3704.	0.01780	4924.54	8.23
120	36.00	5450.	55	3610.	0.02890	11033.54	9.86
121	36.00	7350.	74	3618.	0.01550	11010.43	7.10
122	30.00	5280.	52	3699.	0.01650	9276.51	9.59
123	24.00	2320.	23	3675.	0.02480	4908.38	5.41
124	24.00	720.	7	3747.	0.02480	3947.05	1.09
125	24.00	3200.	32	3643.	0.02480	4579.54	6.50
126	24.00	1990.	20	3625.	0.02120	8325.59	11.42
127	24.00	4000.	39	3736.	0.02120	6270.89	13.02
128	24.00	2700.	27	3643.	0.02120	3699.18	3.06
129	24.00	650.	6	3947.	0.02120	2498.18	0.34
130	24.00	8010.	79	3694.	0.01780	2081.18	2.41
131	24.00	2010.	20	3661.	0.02480	2004.01	0.78
132	24.00	2040.	20	3716.	0.02480	1615.01	0.52
133	36.00	3250.	33	3588.	0.02520	-3370.37	0.48
134	36.00	3940.	39	3680.	0.02520	-1673.68	0.14
135	24.00	2260.	22	3742.	0.01780	4985.71	3.90
136	36.00	5320.	53	3657.	0.01550	6559.21	1.82
137	36.00	5650.	57	3611.	0.01550	7074.04	2.25
138	12.00	5280.	48	4007.	0.02240	-833.04	10.26

139	12.00	5800.	53	3987.	0.03130	1399.00	44.40
140	12.00	10520.	96	3992.	0.02670	1306.00	59.87
141	24.00	13250.	131	3685.	0.01780	778.00	0.56
142	12.00	3270.	30	3971.	0.02240	1818.34	30.26
143	12.00	5330.	49	3963.	0.02240	596.88	5.32
144	12.00	5180.	47	4015.	0.01670	67.00	0.05
145	12.00	1040.	10	3789.	0.03130	1111.47	5.03
146	24.00	4060.	40	3698.	0.01780	10482.82	31.01
147	24.00	7120.	70	3705.	0.01780	1171.12	0.68
148	12.00	2620.	24	3977.	0.02670	-210.88	0.39
149	24.00	5800.	57	3707.	0.01780	778.00	0.24
150	12.00	3830.	35	3986.	0.03130	257.00	0.99
151	12.00	7310.	67	3975.	0.03130	103.21	0.30
152	16.00	7580.	72	3835.	0.02850	965.23	5.97
153	16.00	2480.	23	3928.	0.02850	674.01	0.95
154	12.00	4970.	45	4023.	0.03130	435.89	3.69
155	12.00	1880.	17	4029.	0.03130	185.89	0.25
156	30.00	880.	9	3562.	0.02310	1229.00	0.04
157	20.00	1300.	13	3643.	0.02640	1118.00	0.42
158	30.00	730.	7	3799.	0.02670	6454.60	1.04
159	30.00	4400.	44	3643.	0.02670	5709.38	4.90
160	20.00	2270.	22	3759.	0.02640	806.00	0.38
161	24.00	830.	8	3780.	0.02880	4570.38	1.95
162	24.00	1470.	14	3825.	0.02480	4570.38	2.97
163	12.00	2310.	21	4007.	0.03630	342.23	1.23
164	14.00	380.	4	3461.	0.03450	1111.23	0.94
165	12.00	1020.	9	4129.	0.03630	1111.23	5.71
166	24.00	470.	5	3424.	0.02480	6540.79	1.95
167	30.00	7200.	72	3643.	0.01650	10570.17	16.98
168	14.00	1480.	14	3851.	0.03450	1082.00	3.46
169	24.00	1990.	20	3625.	0.02480	5825.57	6.54
170	12.00	4530.	41	4025.	0.03630	910.82	17.05
171	48.00	1700.	18	3441.	0.01000	-2667.89	0.01
172	16.00	1380.	13	3867.	0.03300	1342.02	2.43
173	16.00	750.	7	3903.	0.03300	1239.02	1.13
174	16.00	1700.	16	3871.	0.03300	1093.02	1.99
175	16.00	1830.	17	3922.	0.02040	815.02	0.74
176	12.00	7920.	72	4007.	0.03130	329.02	3.35
177	12.00	2700.	25	3934.	0.02670	-398.98	1.43
178	12.00	3900.	36	3947.	0.02670	152.35	0.30
179	24.00	1400.	14	3643.	0.01780	-337.26	0.01
180	16.00	2200.	21	3816.	0.02040	-337.26	0.15
181	16.00	2200.	21	3816.	0.02040	-552.26	0.41
182	24.00	1450.	14	3773.	0.01780	4254.60	1.82
183	12.00	8280.	76	3969.	0.03130	104.00	0.35
184	24.00	1500.	15	3643.	0.01780	6236.86	4.06
185	24.00	1500.	15	3643.	0.01780	6125.86	3.91
186	36.00	5900.	59	3643.	0.02890	3236.00	0.92
187	24.00	1000.	10	3643.	0.01780	1881.00	0.25
188	24.00	4860.	48	3689.	0.01780	1618.00	0.88
189	24.00	3390.	33	3742.	0.02120	-243.00	0.02
190	24.00	1500.	15	3643.	0.02480	-924.00	0.12
191	24.00	1000.	10	3643.	0.02480	-1355.00	0.18
192	36.00	5850.	59	3612.	0.02890	-1355.00	0.16
193	16.00	5280.	50	3847.	0.02420	1389.85	7.32
194	16.00	5260.	50	3832.	0.02850	1009.07	4.53
195	16.00	2630.	25	3832.	0.02850	17.10	0.00

196	16.00	2480.	23	3928.	0.02850	674.01	0.95
197	12.00	5300.	48	4022.	0.03130	678.96	9.56
198	12.00	7360.	67	4002.	0.03130	520.55	7.80
199	24.00	5280.	52	3699.	0.02880	2164.72	2.78
200	24.00	2620.	26	3671.	0.02880	357.41	0.04
201	12.00	2240.	20	4080.	0.03630	961.33	9.39
202	12.00	3580.	33	3952.	0.03630	711.33	8.22
203	16.00	1200.	11	3974.	0.02850	1448.15	2.13
204	24.00	400.	4	3643.	0.02880	1316.15	0.08
205	24.00	780.	8	3552.	0.02880	-55.06	0.00
206	24.00	1980.	20	3607.	0.02880	-2194.44	1.07
207	16.00	840.	8	3825.	0.02850	1197.21	1.02
208	16.00	1120.	11	3709.	0.02850	900.12	0.77
209	20.00	410.	4	3734.	0.02640	2028.38	0.43
210	24.00	1970.	19	3777.	0.01780	-696.66	0.07
211	20.00	430.	4	3916.	0.02640	2412.04	0.64
212	16.00	760.	7	3955.	0.02850	297.10	0.06
213	16.00	1950.	18	3947.	0.02850	297.19	0.15
214	20.00	1140.	11	3775.	0.02640	2133.95	1.33
215	14.00	2910.	27	3926.	0.03450	1116.70	7.24
216	16.00	2480.	23	3928.	0.03300	873.70	1.85
217	20.00	770.	7	4007.	0.02640	1773.81	0.62
218	20.00	1460.	14	3799.	0.02640	2017.77	1.52
219	16.00	760.	7	3955.	0.02850	1612.00	1.67
220	16.00	1600.	15	3886.	0.02850	538.05	0.39
221	16.00	3100.	29	3894.	0.03300	747.95	1.70
222	20.00	1550.	15	3764.	0.02640	2285.27	2.08
223	36.00	1500.	15	3643.	0.01550	-1870.69	0.04
224	16.00	560.	5	4080.	0.02040	-1870.69	1.19
225	16.00	5780.	55	3828.	0.02040	-1560.57	8.52
226	12.00	5220.	48	3962.	0.02670	-562.13	5.50
227	16.00	600.	6	3643.	0.02850	1597.98	1.29
228	12.00	1740.	16	3962.	0.03130	1597.98	17.38
229	12.00	3500.	32	3985.	0.03130	864.24	10.22
230	12.00	5300.	48	4022.	0.03130	198.09	0.81
231	16.00	2400.	23	3801.	0.02040	2127.86	6.57
232	16.00	2200.	21	3816.	0.03300	2572.14	14.24
233	12.00	2000.	18	4048.	0.03130	683.26	3.65
234	12.00	3900.	36	3947.	0.03130	247.74	0.94
235	16.00	6800.	64	3871.	0.02200	1548.88	10.64
236	16.00	1800.	17	3857.	0.02200	1263.88	1.88
237	16.00	6000.	57	3835.	0.02040	-687.29	1.71
238	16.00	2000.	19	3835.	0.03300	1075.57	2.26
239	16.00	1360.	13	3811.	0.03300	943.57	1.18
240	16.00	5620.	53	3863.	0.02040	761.09	1.97
241	16.00	1850.	18	3744.	0.02420	592.24	0.47
242	12.00	3090.	28	4020.	0.03130	346.01	1.45
243	12.00	8300.	76	3979.	0.02240	1039.60	25.11
244	48.00	2300.	25	3352.	0.01000	-1227.92	0.00
245	48.00	2400.	26	3363.	0.01000	-188.32	0.00
246	24.00	2770.	27	3737.	0.01780	4036.91	3.14
247	24.00	2700.	27	3643.	0.01780	2247.76	0.95
248	24.00	1270.	13	3559.	0.01780	2073.76	0.38
249	12.00	4930.	45	3991.	0.03130	-13.89	0.00
250	12.00	5380.	49	4000.	0.03130	-458.89	4.43
251	24.00	5340.	53	3670.	0.01780	1518.04	0.86
252	24.00	2170.	21	3764.	0.02480	1407.41	0.42

253	12.00	2340.	21	4059.	0.03130	-306.37	0.86
254	12.00	5270.	48	4000.	0.03130	370.73	2.83
255	12.00	5670.	52	3972.	0.03130	-646.37	9.27
256	36.00	4720.	47	3659.	0.02170	9014.10	4.28
257	12.00	5280.	48	4007.	0.02670	-580.75	5.94
258	24.00	3990.	39	3727.	0.02120	-2174.88	1.56
259	36.00	5920.	59	3655.	0.01550	9846.03	4.57
260	12.00	5380.	49	4000.	0.02670	-492.57	4.36
261	12.00	5420.	50	3949.	0.02670	627.70	7.12
262	12.00	5470.	50	3985.	0.02670	466.43	3.97
263	12.00	5230.	48	3969.	0.02240	-602.31	5.31
264	36.00	5280.	53	3629.	0.01550	14366.22	8.69
265	36.00	4760.	48	3613.	0.01850	13055.91	7.72
266	48.00	800.	8	3643.	0.01680	-64476.84	6.82
267	12.00	5340.	49	3970.	0.02670	555.17	5.49
268	12.00	5200.	48	3947.	0.02240	-110.50	0.18
269	12.00	5240.	48	3977.	0.02240	-200.68	0.59
270	36.00	5280.	53	3629.	0.02170	-3823.20	0.86
271	36.00	3680.	37	3623.	0.02170	-4627.70	0.88
272	36.00	1600.	16	3643.	0.02170	-5134.70	0.47
273	12.00	5280.	48	4007.	0.02240	870.94	11.21
274	36.00	5300.	53	3643.	0.01550	2854.77	0.34
275	12.00	5280.	48	4007.	0.02240	403.75	2.41
276	16.00	3380.	32	3848.	0.02670	2768.49	20.51
277	16.00	6460.	60	3922.	0.02670	2818.24	40.62
278	16.00	3300.	31	3878.	0.02670	1714.24	7.68
279	16.00	2800.	26	3923.	0.02670	-243.00	0.13
280	16.00	5180.	48	3931.	0.02670	-706.17	2.05
281	24.00	5280.	52	3699.	0.01780	464.24	0.08
282	24.00	10500.	104	3678.	0.01780	5310.62	20.58
283	24.00	5400.	53	3712.	0.01780	2754.00	2.85
284	24.00	9000.	89	3684.	0.01780	2754.00	4.74
285	24.00	11000.	108	3710.	0.01780	2608.00	5.20
286	24.00	500.	5	3643.	0.01780	-3801.96	0.50
287	24.00	18400.	181	3703.	0.01780	5326.96	36.29
288	24.00	6100.	60	3704.	0.01780	-6502.84	17.93
289	24.00	4470.	44	3701.	0.01780	-7384.84	16.94
290	24.00	7100.	70	3695.	0.01780	174.00	0.01
291	12.00	7390.	68	3959.	0.02670	-36.83	0.03
292	12.00	7700.	70	4007.	0.02670	-279.83	2.01
293	12.00	5270.	48	4000.	0.02670	698.87	8.59
294	12.00	2690.	25	3920.	0.02670	490.87	2.16
295	12.00	4610.	42	3999.	0.02670	240.87	0.89
296	12.00	5260.	48	3992.	0.02240	523.20	4.03
297	60.00	2700.	36	2732.	0.01000	9237.18	0.09
298	60.00	5300.	71	2719.	0.01000	8883.18	0.17
299	60.00	5100.	68	2732.	0.01000	7850.84	0.13
300	24.00	2630.	26	3685.	0.01780	-503.46	0.05
301	24.00	7050.	70	3669.	0.01780	-656.46	0.21
302	24.00	6070.	60	3686.	0.01780	-1656.46	1.16
303	24.00	5280.	52	3699.	0.01780	2312.00	1.96
304	12.00	7350.	67	3996.	0.02240	1243.00	31.79
305	36.00	5380.	54	3630.	0.02170	1643.06	0.16
306	36.00	5220.	52	3657.	0.02170	222.00	0.00
307	12.00	5570.	51	3979.	0.02240	692.06	7.47
308	12.00	5280.	48	4007.	0.02240	-451.77	3.02
309	60.00	10560.	109	3529.	0.01310	5044.95	0.14

310	60.00	10500.	140	2732.	0.01000	15211.11	0.97
311	60.00	12000.	160	2732.	0.01000	15211.11	1.11
312	48.00	1800.	19	3451.	0.01000	24812.36	1.35
313	48.00	3900.	42	3383.	0.01000	-3435.96	0.06
314	48.00	2300.	25	3352.	0.01000	-13244.81	0.49
315	48.00	3100.	33	3422.	0.01000	-2564.48	0.02
316	48.00	8000.	86	3389.	0.01000	1378.12	0.02
317	48.00	5200.	56	3383.	0.01000	13485.48	1.15
318	36.00	5300.	55	3511.	0.01040	-12934.14	4.74
319	24.00	16600.	166	3643.	0.01200	2258.62	3.97
320	48.00	4000.	43	3389.	0.01000	-864.24	0.00
321	36.00	2150.	22	3560.	0.01850	27793.66	15.80

Model City A (Large Network System)

CONNECTIVITY

LINE	U/S NODE	D/S NODE
1	1	2
2	2	3
3	3	42
4	4	5
5	5	6
6	6	7
7	7	8
8	8	9
9	9	10
10	10	11
11	11	12
12	12	13
13	13	14
14	14	15
15	15	16
16	16	17
17	17	18
18	18	19
19	19	20
20	20	21
21	21	22
22	22	23
23	23	24
24	24	25
25	25	26
26	26	27
27	27	28
28	28	29
29	29	30
30	30	31
31	31	32
32	32	33
33	33	34
34	34	35
35	34	35
36	35	36
37	35	36
38	36	37
39	37	38
40	38	39
41	39	40
42	40	41
43	1	43
44	43	44
45	44	45
46	45	46
47	46	47
48	47	48
49	48	49
50	49	50
51	50	51
52	51	52
53	52	53

54	53	54
55	54	55
56	55	56
57	56	57
58	57	58
59	58	59
60	59	60
61	60	61
62	61	62
63	62	63
64	63	64
65	64	65
66	65	66
67	66	67
68	43	76
69	57	68
70	68	69
71	69	70
72	70	71
73	71	72
74	71	72
75	72	73
76	73	75
77	61	74
78	74	75
79	74	75
80	1	76
81	76	77
82	77	78
83	78	79
84	79	80
85	80	81
86	81	82
87	82	83
88	83	84
89	84	85
90	85	57
91	84	86
92	86	87
93	87	88
94	49	82
95	88	89
96	89	90
97	90	91
98	91	92
99	92	93
100	93	94
101	94	95
102	95	96
103	96	97
104	97	98
105	76	99
106	99	100
107	100	101
108	101	102
109	102	103
110	102	96

111	103	104
112	104	98
113	1	105
114	1	105
115	105	106
116	106	107
117	107	108
118	108	109
119	109	47
120	105	110
121	110	111
122	111	112
123	112	113
124	113	114
125	112	114
126	114	115
127	115	116
128	116	117
129	117	118
130	118	119
131	119	120
132	120	121
133	121	122
134	122	30
135	46	79
136	67	37
137	42	5
138	25	26
139	2	123
140	123	124
141	124	125
142	3	126
143	126	127
144	127	123
145	126	229
146	4	229
147	229	128
148	128	127
149	128	129
150	229	130
151	42	131
152	131	8
153	111	131
154	2	132
155	132	110
156	9	133
157	133	134
158	87	135
159	135	136
160	136	137
161	136	138
162	138	139
163	135	140
164	140	141
165	141	142
166	142	139
167	139	20

282	210	211
283	211	212
284	212	213
285	213	214
286	214	94
287	214	102
288	210	215
289	215	216
290	216	217
291	208	218
292	218	207
293	80	219
294	219	220
295	220	222
296	222	86
297	81	221
298	221	222
299	222	216
300	100	223
301	223	224
302	224	210
303	100	225
304	225	226
305	88	227
306	227	228
307	227	203
308	89	204
309	72	229
310	229	230
311	230	92
312	42	4
313	144	116
314	17	48
315	20	154
316	154	24
317	121	150
318	26	150
319	211	101
320	116	183
321	21	231

APPENDIX H
LISTING OF PARTICIPATING UTILITIES

The following utilities participated in this research study through in-kind contributions

BHC COMPANY
BOARD OF PUBLIC UTILITIES
CALIFORNIA WATER SERVICE COMPANY
CORCORAN, CITY OF
DES MOINES WATER WORKS
FREDERICTON, CITY OF
INGLEWOOD, CITY OF
IOWA, CITY OF
LAS VEGAS VALLEY WATER DISTRICT
LA VERNE, CITY OF
MESA, CITY OF
METROPOLITAN WATER DISTRICT OF SOUTHERN CALIFORNIA
NORTH LAS VEGAS, CITY OF
NORTHGLENN, CITY OF
OLYMPIA, CITY OF
PERU, CITY OF
SAN DIEGO, CITY OF
SOUTHERN CALIFORNIA WATER COMPANY
SWEETWATER AUTHORITY
TOWN OF CASTLE ROCK, TOWN OF
WALNUT VALLEY WATER DISTRICT
WATER DIST. NO. 1 OF JOHNSON COUNTY

APPENDIX I
LISTING OF RESPONDENTS

The following responded to the initial survey and/or the follow-up survey of this project

ADRIAN WATER DEPARTMENT
ALAMEDA COUNTY WATER DISTRICT
ALLENDALE TOWNSHIP WATER
ALPENA TOWNSHIP DPW
ALPINE WATER USERS ASSOCIATION
AMERICAN COMMONWEALTH MANAGEME
AMERICAN WATER RESOURCES
AMHERST COUNTY SERVICE AUTHORI
ANN ARBOR UTILITIES DEPARTMENT
APPOMATTOX RIVER WATER AUTH
ARIZONA WATER COMPANY
ARLINGTON WATER DEPARTMENT
ARVIN COMMUNITY SERVICES DISTRICT
ASSUMPTION PARISH WATER WORKS
ATHENS UTILITIES BOARD
AUBURN DEPARTMENT OF PUBLIC WO
AUSTIN UTILITIES
AVION WATER COMPANY INCORPORAT
BAA HARRISBURG
BARABOO CITY WATER WORKS
BARNSTABLE WATER DEPARTMENT
BASS LAKE WATER COMPANY
BATAVIA BUREAU WATER & SEWER
BATH WATER DISTRICT
BATTLEMENT MESA METROPOLITAN D
BAYOU LIBERTY WATER ASSOCIATIO
BEATRICE PUBLIC WORKS BOARD
BEAUFORT JASPER WATER AUTH
BECKER JIBA WATER SUPPLY CORPO
BELL COUNTY WATER CONTROL 1
BELLA VISTA WATER COMPANY INCO
BELLA VISTA WATER DISTRICT
BELVIDERE WATER DEPARTMENT
BENBROOK WATER & SEWER AUTHORI
BEREA COLLEGE UTILTIES
BERLIN WATER CONTROL COMM
BETHPAGE WATER DISTRICT
BHC COMPANY
BLOOMFIELD TOWNSHIP
BLUE LAKE SPRINGS MUTUAL WATER
BLUMFIELD REESE WATER AUTHORIT

BOARD OF PUBLIC UTILITIES
BODEGA BAY PUBLIC UTILITY DISTRICT
BOISTFORT VALLEY WATER
BOONVILLE MUNICIPAL COMM
BOURNE WATER DISTRICT
BRIDGEPORT CHARTER TOWNSHIP
BRIGHAM CITY CORPORATION
BROADVIEW PARK WATER COMPANY
BROADVIEW TERRACE MUTUAL WATER
BROOKFIELD WATER UTILITY BD
BROOKTRAILS COMMUNITY SERVICES
BRUNSWICK & TOPSHAM WATER DISTRICT
BRUNSWICK COUNTY PUBLIC UTILIT
BRYAN COMPANY RURAL WATER DIST
BRYN MAWR LAKERIDGE WATER & SE
BUCKS COUNTY WATER & SEWER AUT
BURTCHVILLE TWP D P W -WATER
BUZZARDS BAY WATER DISTRICT
CALAVERAS COUNTY WATER DIST
CALAVERAS PUBLIC UTIL DIST
CALLAYOMI COUNTY WATER DIST
CALLEGUAS MUNICIPAL WATER DIST
CAMP NAVAJO
CANTON BOROUGH AUTHORITY
CARIBOU WATERWORKS
CARLE PLACE WATER DISTRICT
CARMEL WATER DEPARTMENT
CARROLL COUNTY WATER DISTRICT
CARSON CITY WATER UTILITY DIV.
CARTHAGE WATER & ELECTRIC
CASS COUNTY PWSD #10
CASSELBERRY PUBLIC WORKS DEPT
CEDAR HILL PUBLIC WORKS
CEDAR RAPIDS WATER DEPARTMENT
CEDAR RIVER WATER AND SEWER DI
CEDAR-KNOX RURAL WATER PROJECT
CEDARVILLE WATER USERS ASSOCIA
CELINA WATER PLANT
CENTENNIAL WATER & SANITATION
CENTER MUNICIPAL WATER SYSTEM
CENTRAL STICKNEY SANITARY
CENTRAL TEXAS WSC
CHARLES CITY COUNTY P W D
CHARLESTON COMM OF PUBLIC WORK
CHARLOTTE HARBOR WATER ASSN
CHARTER TOWNSHIP OF BREITUNG

CHELAN COUNTY PUD #1
CHENOWITH WATER PUD
CHESTER WATER AUTHORITY
CITRUS HEIGHTS WATER DISTRICT
CITY & COUNTY OF HONOLULU
CITY AND BOROUGH OF SITKA
CITY OF ADELANTO
CITY OF AIRWAY HEIGHTS
CITY OF AKRON
CITY OF ALAMOSA
CITY OF ALGONA
CITY OF ALPENA ETOS
CITY OF ALTOONA
CITY OF AMES
CITY OF ANAHEIM
CITY OF ARCADIA
CITY OF ASPEN
CITY OF AUSTIN
CITY OF BAKER CITY
CITY OF BANKS
CITY OF BARRIE
CITY OF BEMIDJI
CITY OF BENICIA
CITY OF BISHOP
CITY OF BLANDING
CITY OF BLYTHE WATER
CITY OF BOONVILLE PUBLIC WORKS
CITY OF BRENTWOOD
CITY OF BROOKFIELD
CITY OF BRYANT WATER & WASTEWA
CITY OF BUENA PARK
CITY OF BURBANK
CITY OF BURKESVILLE WATERWORKS
CITY OF BURLESON
CITY OF BURLINGTON
CITY OF CALGARY WATERWORKS
CITY OF CALIFORNIA CITY
CITY OF CAMARILLO
CITY OF CAMROSE
CITY OF CARROLLTON
CITY OF CASPER PUBLIC UTILITIE
CITY OF CAVE JUNCTION
CITY OF CERES
CITY OF CHESAPEAKE UTILITIES
CITY OF COLLEGE STATION
CITY OF COLUMBIA FALLS

CITY OF COLUSA
CITY OF CORCORAN
CITY OF DAVIS
CITY OF DECATUR
CITY OF DEERFIELD BEACH
CITY OF DEL MAR
CITY OF E CHICAGO
CITY OF EDEN PRAIRIE WATER DEP
CITY OF EDMONDS
CITY OF EL CENTRO
CITY OF EL PASO
CITY OF ELGIN WATER DEPARTMENT
CITY OF ESCONDIDO
CITY OF EVANS
CITY OF EVART
CITY OF EXCELSIOR SPRINGS
CITY OF FAIRFIELD
CITY OF FAIRVIEW
CITY OF FLAGSTAFF
CITY OF FLINT
CITY OF FLORENCE
CITY OF FORT BRAGG
CITY OF FORT COLLINS WATER UTILITIES
CITY OF FOUNTAIN VALLEY WATER DEPT.
CITY OF FREDERICTON
CITY OF FULLERTON
CITY OF GALENA PARK
CITY OF GALION
CITY OF GARLAND
CITY OF GEORGE
CITY OF GERING
CITY OF GLENDALE
CITY OF GLENDORA
CITY OF GOLDEN PW / WTP
CITY OF GONZALES
CITY OF HANFORD
CITY OF HEMET
CITY OF HILLSBORO
CITY OF HUBBARD
CITY OF INGLEWOOD
CITY OF IRRIGON
CITY OF IRVING
CITY OF JACKSON
CITY OF JOLIET
CITY OF JUNEAU WATER DEPARTMEN
CITY OF KEARNEY

CITY OF KEIZER
CITY OF KELOWNA
CITY OF KINGSFORD
CITY OF LA HABRA
CITY OF LACEY
CITY OF LAMAR
CITY OF LAVERNE
CITY OF LEAVENWORTH
CITY OF LEEDS WATERWORKS
CITY OF LOMA LINDA
CITY OF LOMITA
CITY OF LONG CREEK
CITY OF LUVERNE
CITY OF MANTECA
CITY OF MERCED
CITY OF MESA
CITY OF MIDLAND
CITY OF MILPITAS
CITY OF MINOT
CITY OF MIRAMAR
CITY OF MISSION
CITY OF MONCTON
CITY OF MONROVIA
CITY OF MONTEREY PARK
CITY OF MOOSE JAW
CITY OF MOSES LAKE
CITY OF MT PLEASANT
CITY OF NASHVILLE
CITY OF NEWPORT
CITY OF NEWPORT BEACH UTILITIE
CITY OF NORTH LAS VEGAS
CITY OF NORTHVILLE PUBLIC WORK
CITY OF NORWICH
CITY OF OLDSMAR
CITY OF OLYMPIA
CITY OF ONAWA
CITY OF ORANGEBURG DPU
CITY OF OSAGE CITY
CITY OF OTTAWA
CITY OF OVIEDO
CITY OF OXNARD
CITY OF PEORIA
CITY OF PERU
CITY OF PHILADELPHIA
CITY OF PHOENIX, OREGON
CITY OF PLAINVIEW

CITY OF PLANTATION UTILITIES
CITY OF PORT ST LUCIE UTILITY
CITY OF PORTSMOUTH WATER WORKS
CITY OF POST FALLS
CITY OF PRAIRIE CITY
CITY OF PULASKI WATER & SEWER
CITY OF RADFORD
CITY OF RAINIER
CITY OF RAPID CITY
CITY OF REDLANDS
CITY OF REDMOND
CITY OF REVELSTOKE
CITY OF RIALTO
CITY OF RICHLAND
CITY OF RICHLAND HILLS
CITY OF RICHMOND
CITY OF RIDGELAND
CITY OF RIPON
CITY OF ROSEVILLE
CITY OF SACRAMENTO
CITY OF SAGINAW
CITY OF SALEM, VA
CITY OF SALISBURY
CITY OF SALMON
CITY OF SAN CLEMENTE
CITY OF SAN DIEGO
CITY OF SAN JOSE
CITY OF SANDPOINT
CITY OF SANTA BARBARA
CITY OF SANTA CLARA
CITY OF SANTA CRUZ
CITY OF SANTA FE SPRINGS
CITY OF SANTA ROSA
CITY OF SAVANNAH
CITY OF SCOTTSDALE
CITY OF SEBRING
CITY OF SENATOBIA
CITY OF SENECA
CITY OF SERGEANT BLUFF IOWA
CITY OF SHAFTER
CITY OF SHERIDAN
CITY OF SHERWOOD
CITY OF SOLVANG
CITY OF ST HELENS
CITY OF ST IGNACE
CITY OF ST MARYS

CITY OF STAYTON
CITY OF STERLING
CITY OF TALENT
CITY OF TEMPE
CITY OF THORNTON
CITY OF THUNDER BAY
CITY OF TROUTDALE
CITY OF TUKWILA
CITY OF TULARE
CITY OF TUMWATER
CITY OF TURLOCK
CITY OF UKIAH
CITY OF VALENTINE
CITY OF VENETA
CITY OF VERNON
CITY OF VERNONIA
CITY OF WACONIA
CITY OF WALHALLA
CITY OF WARREN
CITY OF WASHOUGAL
CITY OF WEST PALM BEACH
CITY OF WESTPORT
CITY OF WILSONVILLE
CITY OF WINNIPEG WATER
CITY OF WINSLOW
CITY OF WINTERS
CITY OF WYOMING WATER DEPARTME
CITY OF YACHATS
CITY OF YELM
CLALLAM COUNTY PUD #1
CLEARFIELD MUNIC AUTHORITY
CLEARWATER PUBLIC UTILITIES
CLEVELAND DIVISION OF WATER
COBB-MARIETTA WATER AUTHORITY
COLEBROOK WATER WORKS
COLLIER COUNTY PUBLIC WORKS
COLUMBUS WATER WORKS
COMMONFIELDS OF CAHOKIA
COMMUNITY WATER SYSTEM
CONCORD WATER USERS ASSOCIATIO
CONEMAUGH TOWNSHIP MUNICIPAL A
CORTLAND WATER BOARD
COTTONWOOD WATER WORKS
COUNTY OF GLOUCESTER
COUNTY OF LANDER
COUNTY OF ROANOKE WATER DISTRI

COVINGTON WATER DISTRICT
COWETA COUNTY WATER DEPARTMENT
CRANBERRY MUNICIPAL AUTHORITY
CRESTLINE-LAKE ARROWHEAD AGENC
CRESTVIEW WATER & SANITATION D
CRYSTAL SPRINGS WATER DISTRICT
CUCAMONGA COUNTY WATER DISTRIC
D C WATER & SEWER AUTHORITY
DAYTON WATER DEPARTMENT
DECATUR WATER UTILITY
DEKALB COUNTY PUBLIC WORKS
DEL PASO MANOR COUNTY WATER
DEL-COMPANY WATER COMPANY INCO
DELHI COUNTY WATER DISTRICT
DELTA CHARTER TOWNSHIP
DENVER WATER DEPARTMENT
DEPT WATER & WASTE MANAGEMENT
DES MOINES WATER WORKS
DESERT WATER AGENCY
DIST OF OAK BAY
DIX HILLS WATER DISTRICT
DOVER WATER COMPANY INCORPORAT
DOVER WATER DEPARTMENT
DUBLIN SAN RAMON SERVICE DIS
E PETERSBURG BORO
EAGLE GROVE WATER SUPPLY
EAGLE RIVER WATER & SANITATION
EAST BAY MUNICIPAL UTILITY DIS
EAST ORANGE COUNTY WD
EASTERN MUNICIPAL WATER DISTRI
EASTERN WASHINGTON UNIVERSITY
EL PRADO WATER & SANITATION DI
ELWOOD WATER AND SEWAGE
ENDICOTT MUNICIPAL WATER
ENFIELD DEPT OF PUBLIC WORKS
ENTRANOSA WATER COMPANY-OP ASS
ERIE HURON CNTY RURAL WTR AUTH
ERIE WATER DEPARTMENT
ERWIN UTILITIES
ESCANABA MUNICIPAL WATER DEPT
ESSEX DEPT OF PUBLIC WORKS
EUGENE WATER ELECTRIC BOARD
EVANSVILLE WATER & SEWER DEPAR
FACILITIES MAINTANANCE DEPARTM
FAIR OAKS WATER DISTRICT
FAIRFIELD COUNTY UTILITIES

FLORIDA KEYS AQUEDUCT AUTHORIT
FORESTHILL PUBLIC UTILITY DIST
FREEPORT VILLAGE WATER DEPARTM
FREEPORT WATER & SEWER COMMISS
FREMONT DEPARTMENT OF UTILITIES
FRESNO CITY WATER DIVISION
FT RECOVERY WATER DEPARTMENT
GAFFNEY BOARD OF PUBLIC WORKS
GARDINER WATER DISTRICT
GEAUGA COUNTY DEPARTMENT OF WA
GEORGETOWN COUNTY WATER & SEWE
GEORGETOWN WATER DEPARTMENT
GERMANTOWN WATER UTILITY
GLACIER SPRINGS WATER ASSOCIAT
GLEN COVE WATER DEPARTMENT
GLENSHIRE MUTUAL WATER COMPANY
GLENWOOD WATER WORKS
GRAND ISLAND WATER & LIGHT
GRAND RAPIDS WATER DEPARTMENT
GRAND TRAVERSE COUNTY
GREATER NANAIMO WATER DIST
GREATER VANCOUVER WATER DIST
GREEN VALLEY SPECIAL UTILITY D
GREENEVILLE WATER AND LIGHT
GRENELEFE UTILITIES
GRESHAM WATER & SEWER UTILITY
GREY FOREST WATER SYSTEM
GROTON WATER DEPARTMENT
GROVELAND COMMUNITY SERVICE DI
HALEDON WATER DEPARTMENT
HAMILTON WESTERN UTILITY INCOR
HANOVER COUNTY UTILITIES
HANOVER WATER WORKS COMPANY
HARRIS COUNTY WATER CONTROL
HELLERTOWN BOROUGH AUTHORITY
HESPERIA WATER DISTRICT
HIGH VIEW WATER DISTRICT
HIGHLAND COUNTY WATER COMPANY
HIGHLANDS WATER COMPANY
HOLLAND CHARTER TOWNSHIP
HOLLIDAY WATER COMPANY
HULEN MEADOWS WATER COMPANY
HUMBOLDT COMMUNITY SERVICE
IMPERIAL PUBLIC WATER
INCLINE VILLAGE GEN. IMPROVEMENT DIST.
INDEPENDENCE WATER DEPARTMENT

INDIANA AMERICAN WATER COMPANY
IOWA AMERICAN WATER COMPANY
IRVIN WATER DISTRICT #6
JEFFERSON COUNTY #1 PUBLIC WAT
JEFFERSON COUNTY WATER CONTROL
JENNINGS NORTHWEST REGION UTIL
JOHNSON CITY WATER DEPARTMENT
JORDAN VALLEY WATER CONSERVANC
JORNADA WATER COMPANY INCORPOR
JUNE LAKE PUBLIC UTILITY DISTR
KACHINA VILLAGE IMPROVEMENT DI
KENNETT SQUARE WATER DEPARTMEN
KENTUCKY-AMERICAN WATER COMPAN
KERN COUNTY WATER AGENCY
KETCHIKAN PUBLIC UTILITY
KING COUNTY WATER DIST #119
KING GEORGE COUNTY SERVICE AUT
KINGSTON WATER DISTRICT
KISSIMMEE WATER & SEWER DEPT
KNOXVILLE UTILITIES BOARD
KOLIN RUBY WISE WATERWORKS
L A WATER COOPERATIVE
LA PORTE CITY WATER DEPARTMENT
LA PUENTE VALLEY WATER DISTRIC
LACEY MUNIC UTILITY AUTHORITY
LAFAYETTE UTILITIES SYSTEM
LAGOON ESTATES WATER COMPANY I
LAKE DON PEDRO COMMUNITY
LAKE ELIZABETH MUTUAL WATER COMPANY
LAKE HAVASU CITY
LAKE KILBY WTP
LAKE PARK MUNICIPAL UTILITIES
LAKE TOWNSHIP WATER SYSTEM
LAKELAND WATER UTILITY
LAKESIDE COUNTY WATER & SEWER
LAMONI MUNICIPAL UTILITIES
LANCASTER BUREAU OF WATER
LANCASTER OHIO WATER DEPARTMENT
LARAMIE WATER DEPARTMENT
LAS VEGAS VALLEY WATER DISTRIC
LATHAM WATER DISTRICT
LAURENS COUNTY WATER & SEWER
LAWRENCE WATER & SEWER DEPARTM
LEES SUMMIT WATER UTILITIES DE
LEHIGH COUNTY AUTHORITY
LINDA COUNTY WATER DISTRICT

LITCHFIELD WATER DEPARTMENT
LITTLE RIVER W & S INCORPORATE
LITTLETON WATER & LIGHT DEPT
LONG BEACH WATER DEPARTMENT
LOS ALAMOS COUNTY UTILITIES
LOUISVILLE WATER COMPANY
LOWELL REGIONAL WATER UTILITY
LOWER LAKE COUNTY WATER WORKS
LOWER SAUCON AUTHORITY
LYNDEN WATER DEPARTMENT
MACON COUNTY WATER AUTHORITY
MADISON SUBURBAN UTILITY
MAGNA WATER COMPANY
MAMMOTH COMMUNITY WATER DISTRI
MANATEE COUNTY PUBLIC UTILITIES DEPT.
MANCHESTER WATER DEPARTMENT
MANCHESTER WATER WORKS
MARIN MUNICIPAL WATER DISTRICT
MARION WATER DEPARTMENT
MARSHALL MUNICIPAL UTILITY
MARSHALL WATER WORKS INCORPORA
MARSHALLTOWN WATER WORKS
MAYWOOD MUTUAL WATER COMPANY #
MCFARLAND MUTUAL WATER COMPANY
METROPOLITAN UTILITIES DISTRIC
METROPOLITAN WATER DISTRICT - GLENDORA OFFICE
METROPOLITAN WATER WORKS
MIAMI DADE WATER & SEWER DEPAR
MIDLAND WATER DEPARTMENT
MILAN DEPARTMENT OF PUBLIC UTI
MILITARY HIGHWAY WSC
MILWAUKEE WATER WORKS
MISSION HILLS CSD
MISSION SPRINGS WATER DISTRICT
MOAB IRRIGATION DISTRICT #20
MOAPA VALLEY WATER DIST.
MODEL IRRIGATION DIST #18
MOLINE DEPARTMENT OF PUBLIC UT
MONTE DEL LAGO MOBILE HOMES
MOUNT PLEASANT WATER WORKS
MOUNTAIN TOP WATER SYSTEM
MT LEMMON COOP WATER COMPANY
MUSKINGUM CONSERVANCY DIST
MYOMA DUNES WATER COMPANY
N CHELMSFORD WATER DISTRICT
N GUALALA WATER COMPANY

N LAS VEGAS UTILITY
N MARIN WATER DISTRICT
N TABLE MOUNTAIN WATER & SANIT
N TONAWANDA WATER DEPARTMENT
NASHVILLE METRO WATER & SEWER
NEBRASKA CITY UTILITIES
NESKOWIN REGIONAL WATER DIST
NEW LONDON-SPRINGFIELD WATER
NEW MEXICO-AMERICAN WATER COMP
NEW RICHMOND CITY UTILITY
NEWPORT NEWS PUBLIC UTILITIES
NEWTON WATER WORKS
NORTH OF RIVER MUNICIPAL
NORTH SYDNEY WATER UTILITY
NORTHERN ILLINOIS WATER
NORTHFIELD MOUNT HERMAN SCHOOL
NORTHRIDGE WATER DISTRICT
NORTHSHORE UTILITY DISTRICT
NORTHVILLE TOWNSHIP WATER & SE
NORTHWEST WATER COMMISSION
NYC DEPT OF ENVIRONMENTAL PROT
OCONTO FALLS WATER & LIGHT COM
OGDEN CITY DEPT WATER UTILITY
OKALOOSA COUNTY WATER AND SEWE
OLD HICKORY UTILITY DISTRICT
OLD TOWN WATER DISTRICT
OLEAN WATER DEPARTMENT
OLIVENHAIN MUNICIPAL WATER DIS
ONEIDA UTILITIES COMMISSION
ORCHARD DALE WATER DISTRICT
OROVILLE WYANDOTTE IRRIG DIST
OTTER LAKE WATER COMMISSION
OWENSBORO MUNIC UTILITIES
PADUCAH WATER WORKS
PALMDALE WATER DISTRICT
PARADISE IRRIGATION DISTRICT
PARADISE VALLEY WATER COMPANY
PARIS UTILITY DISTRICT
PARK WATER CO. & SUBSIDIARIES
PAWTUCKET WATER SUPPLY BOARD
PENACOOK-BOSCAWEN WATER PRECIN
PENN YAN MUNICIPAL BOARD
PIERCE UTILITIES
PINEDALE COUNTY WATER DISTRICT
PLACER COUNTY WATER AGENCY
PLEASANT GROVE CITY WATER

POMPTON LAKES MUNICIPAL UTILIT
PORT OF MORROW
PORTLAND WATER BUREAU
POWELL VALLEY WATER DISTRICT
PRINCEVILLE UTILITIES COMPANY
PROVINCETOWN WATER DEPARTMENT
PUBLIC WORKS COMMISSION
RAINBOW WATER COMPANY INCORPOR
RANCHO CALIFORNIA WATER DISTRI
RANCHO MURIETA CSD
REDWOOD CITY WATER DEPARTMENT
RINCON RANCH ESTATES WATER
RIVER FALLS MUNICIPAL UTILITY
RIVERVIEW WATER DISTRICT
RIVIERA UTILITY FOLEY UTILITY
ROBINSON TOWNSHIP AUTHORITY
ROCK RAPIDS UTILITIES DIVISION
ROSS COUNTY WATER COMPANY
ROXBORO WATER DEPARTMENT
RUBIO CANON LAND & WATER ASSOC
RUTLAND WATERWORKS
S ADAMS COUNTY WATER SANITATIO
S FARMINGDALE WATER DISTRICT
S HOUSTON WATER/WASTEWATER
S TAWAKONI WATER SUPPLY CORPOR
SALT LAKE CITY WATER DEPT
SAMMAMISH PLATEAU WATER/SEWER
SAN ANTONIO WATER COMPANY
SAN JUAN SUB WATER DISTRICT
SANDWICH WATER DISTRICT
SANTA MARGARITA WATER DISTRICT
SANTA MONICA WATER DIVISION
SANTEE UTILITIES INCORPORATED
SANTIAGO COUNTY WATER DISTRICT
SCHUYLER DEPT OF UTILITIES
SEACOAST UTILITY AUTHORITY
SEASIDE WATER DEPARTMENT
SENECA LIGHT & WATER PLANT
SHORELINE WATER DISTRICT
SIERRA PACIFIC POWER CO. - WESTPAC UTIL.
SKAGIT COUNTY PUD #1
SOLANO IRRIGATION DISTRICT
SONOMA COUNTY WATER AGENCY
SOUTH BAY IRRIGATION DISTRICT
SOUTH BEND WATER WORKS
SOUTH MESA WATER CO.

SOUTH TEXAS WATER AUTHORITY
SPENCER MUNICIPAL UTILITY
SPRINGFIELD UTILITY BOARD
SPRINGFIELD WATER LIGHT & POWE
SQUAW VALLEY PUBLIC SERVICE DI
ST CHARLES MESA WATER DISTRICT
ST DAVID WATER ASSOCIATION
ST GEORGE CITY CORPORATION
ST MARY WATER & SEWER COMM #1
ST MARY'S COUNTY METRO COMM
STAFFORD MUNICIPAL UTILITIES A
STAR MINDINGALL WATER AUTHORIT
STINSON BEACH COUNTY WATER DIS
STOCKTON MUNICIPAL UTILITIES
STOUGHTON WATER UTILITY
STRATMOOR HILLS WATER DISTRICT
STROUDSBURG MUNICIPAL AUTHORIT
SUN VALLEY GENERAL IMPR DIST
SUNNYSLOPE COUNTY WATER DISTRI
SUPERIOR TOWNSHIP UTIL DEPT
T M RURAL WATER DISTRICT
TAMPA BAY WATER
TAMPA WATER DEPARTMENT
TCW SUPPLY INCORPORATED
TELFORD BOROUGH AUTHORITY
TENNESSEE-AMERICAN WATER COMPA
THE CORPORATION OF CITY OF BRA
THE EMPIRE DIST ELECTRIC COMPA
THERMALITO IRRIGATION DISTRICT
THIEF RIVER FALLS WATER
TILLAMOOK WATER COMMISSION
TOWN OF BASALT
TOWN OF BELLEAIR
TOWN OF BENNINGTON WATER
TOWN OF BRASELTON
TOWN OF CARBONDALE
TOWN OF CASTLE ROCK
TOWN OF CHENANGO
TOWN OF GUILDERLAND
TOWN OF HATFIELD WATER DEPARTM
TOWN OF HILLSBOROUGH
TOWN OF MEDLEY
TOWN OF MINDEN
TOWN OF NAGS HEAD
TOWN OF NEDERLAND
TOWN OF NEW CASTLE WD #1

TOWN OF NUCLA
TOWN OF PARKER
TOWN OF PINEDALE
TOWN OF PLAIN DEALING
TOWN OF PLATTEVILLE
TOWN OF PURCELLVILLE
TOWN OF SACKVILLE
TOWN OF SNOWFLAKE
TOWN OF ST JOHN WATER DEPARTME
TOWN OF STEILACOOM
TOWN OF TAYLOR
TOWN OF TELLURIDE
TOWN OF TOPSAIL BEACH
TOWN OF WARRENTON
TOWNSEND WATER DEPARTMENT
TRACT 349 MUTUAL WATER CO., INC.
TRI COUNTY WATER ASSOCIATION
TRIUNFO SANITATION DISTRICT
TWENTYNINE PALMS WATER DISTRIC
TWINING WATER & SANITATION DIS
UNITED WATER CONNECTICUT
UPPER MOHAWK VALLEY REGIONAL W
VALENCIA HEIGHTS WATER COMPANY
VALLECITOS WATER DISTRICT
VALLEY WATER SYSTEMS
VERA WATER & POWER
VILLAGE DE MEMRAMCOOK
VILLAGE OF ADA
VILLAGE OF ARCHBOLD WATER DEPT
VILLAGE OF BETHANY WATER DEPT
VILLAGE OF BIRCH RUN
VILLAGE OF CASEVILLE
VILLAGE OF CHASE
VILLAGE OF CISSNA PARK
VILLAGE OF CLARENDON HILLS
VILLAGE OF CLINTON
VILLAGE OF FOX POINT
VILLAGE OF FRAZEYSBURG
VILLAGE OF LISLE
VILLAGE OF LYNDONVILLE
VILLAGE OF MARATHON UTILITY
VILLAGE OF MCCONNELSVILLE
VILLAGE OF MCGRAW
VILLAGE OF MINEOLA
VILLAGE OF PEMBERVILLE
VILLAGE OF PLAIN CITY

VILLAGE OF ROSCOMMON
VILLAGE OF S ELGIN WATER
VILLAGE OF ST PARIS
VILLAGE OF WASHINGTONVILLE
VILLAGE OF WESTMONT
W DES MOINES WATER WORKS
W SAN BERNARDINO COUNTY WATER
WALLA WALLA WATER DIVISION
WALNUT VALLEY WATER DISTRICT
WARMINSTER TOWNSHIP MUNICIPAL
WASHINGTON TERRACE CITY
WATER DEVELOPMENT COMPANY
WATER DIST. #1 OF JOHNSON COUNTY
WATERLOO WATER & LIGHT COMMISS
WAUKESHA WATER UTILITY
WAUWATOSA WATER DEPARTMENT
WESTERN BERKS WATER AUTHORITY
WESTHAVEN COMMUNITY SERVICES D
WHEELING WATER DEPARTMENT
WHITE ROCK UTILITIES LIMITED
WHITEHALL TOWNSHIP AUTHORITY
WIMBERLEY WATER SUPPLY CORPORA
WINDSOR UTILITIES COMMISSION
WINNETKA WATER & ELECTRIC
WINTHROP UTILITIES DISTRICT
WOODHAVEN WATER COMPANY
WOODINVILLE WATER DISTRICT
WOODMOOR WATER & SANITATION
YORBA LINDA WATER DISTRICT
YORK WATER COMPANY
YORKTOWN CONSOLIDATED WATER DI
YOSEMITE NATIONAL PARK
YUIMA MUNICIPAL WATER DISTRICT
ZONE 7 WATER AGENCY

REFERENCES

ABPA. 1999. *Survey of State and Public Water System Cross-Connection Control Programs.* American Backflow Prevention Association. P. 4

ASME A112.1.2 - 1991 *Air Gaps in Plumbing Systems*

ASSE Standard No. 1011. 1993. *Performance Requirements for Hose Connection Vacuum Breakers.* American Society of Sanitary Engineering.

ASSE Standard No. 1024. 1994. *Performance Requirements for Dual Check Valve Type Backflow Preventers.* American Society of Sanitary Engineering.

ASSE Standard No. 1012. 1993. *Performance Requirements for Backflow Preventer with Intermediate Atmospheric Vent.* American Society of Sanitary Engineering.

CDC. 1998. "Surveillance for Waterborne-Disease Outbreaks – United States, 1995-1996." *Morbidity and Mortality Weekly Report 47.* p. 3.

CDC. 2000. "Surveillance for Waterborne-Disease Outbreaks – United States, 1997-1998." *Morbidity and Mortality Weekly Report 49.* p. 12.

Chaudhry, H. M. 1988. *Applied Hydraulic Transients.* New York: Van Nostrand Reinhold.

Craun, Gunther F. 1986. *Waterborne Diseases in the United States.* CRC Press, Inc. pp. 131, 134.

Islam, M. R. & Chaudhry, M. H. 1998. "Modeling of constituent transport in unsteady flows in pipe networks." *Journal of Hydraulic Engineering, 124.* pp. 1115-1124.

Karney, B. W. & Mcinnis, D. 1990. "Efficient calculation of transient flow in simple pipe networks." *Journal of Hydraulic Engineering, 118.* pp. 1014-1031.

Karney, B. W. & McInnis, D. 1990. "Transient Analysis of Water Distribution Systems." *J. AWWA.* pp. 82, 62-70.

Pezzinga, G. 1999. "Quasi-2D model for unsteady flow in pipe networks." *Journal of Hydraulic Engineering. ASCE.* pp. 125, 676-685.

USC. 1993. *Manual of Cross-Connection Control 9th Edition.* University of Southern California.

US Federal Register. 2000. *Volume 65.* pp. 83015-83024.

IWA Publishing's authorised EU representative for General Product Safety Regulations is Diane D'Arras, 15 rue Duret, 75116 Paris, France, e-mail: safety@iwap.co.uk.

Printed and bound by CPI Group (UK) Ltd, Croydon, CR0 4YY
17/09/2026
02218813-0003